The Reverse Garbage Mulch Book

By the same author:

The Reverse Garbage Garden

the Reverse Garbage Mulch Book

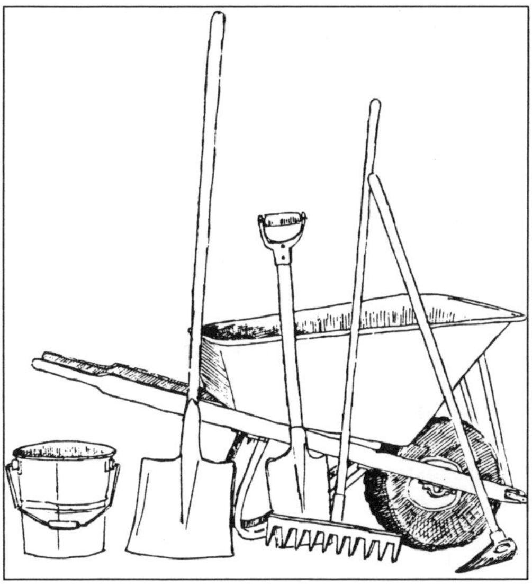

Sandra Clayton

HYLAND HOUSE

First published in 1994 by
Hyland House Publishing Pty Limited
Hyland House
387-389 Clarendon Street
South Melbourne
Victoria 3205

National Library of Australia
Cataloguing-in-publication data

Clayton, Sandra, 1951–
The reverse garbage mulch book.

Bibliography.
Includes index.
ISBN 1 875657 40 1.

1. Mulching. 2. Organic gardening. 3 Recycling (Waste, etc.).
I. Title.

631.451

Typeset by Solo Typesetting, South Australia
Printed in Singapore by Singapore National Printers

Contents

QUICK MULCH REFERENCE

Uses	*Recommended Mulch*	*Comments*
Sources of:		
Nitrogen	alfalfa hay, blood and bone, coffee grounds, compost, feathers, fish wastes, fur, grass clippings, green weeds, hair, human body wastes, leather scraps, leaves, legumes, manures, meat scraps, nut shells, sewage sludge, urine, wool	Most high-nitrogen material is best used with high-bulk mulches (see Coarser and Very Coarse, below). Most decay quickly, but not all.
Phosphorus	blood and bone, bones, grass clippings, hulls, leaf mould and leaves, manures, slag, wood ashes	Mixing your materials gives a wide spread of minerals.
Sulphur	humus (sulphur is made available by the decomposition of organic matter)	Mulching eventually makes humus available.
Potassium	blood and bone, bracken, compost, fish wastes, flue dust from cement works, grass clippings, hay, hulls, leaf mould and leaves, manures, plant wastes, seaweed (and shells), wood ashes	Potassium helps protect plants from disease and aids photosynthesis.
Calcium	bone, ground limestone, leaf mould and leaves, seashells, slag, wood ashes	Acid soils are normally deficient in calcium.
Trace elements	bones, compost, kitchen wastes, leaf mould and leaves, manures, mushroom waste, plant wastes, seaweed, slag, water weeds, weeds, wood ashes, worm castings	Trace elements are vital for healthy plants, but nobody knows how much of each element is needed or how they interact.
Acid mulches	pine needles, leaf mould and leaves	Use only around acid-loving plants or sprinkle with lime or wood ashes.
Alkaline mulches	limestone, seashells, slag, wood ashes	Use sparingly on acid soils, with mulches or in compost.

Uses	*Recommended Mulch*	*Comments*
Mulch for seeds	compost, leaf mould, peat moss, sawdust (rotted), worm castings	No more than 2 centimetres deep and remove any lumps
Mulch for seedlings	compost, grass clippings, leaf mould, mushroom waste, peat moss, sawdust, seashells, worm castings	Use 2–4 centimetres deep and don't let them touch the seedlings.
Coarse mulch for larger seedlings and plants	hay, hair, hulls, kitchen wastes, leaf mould, leaves, manures, mushroom waste, nut shells, packing materials (small or well shredded), peat moss, pine needles, plant wastes, seaweed, shredded bark, silage, straw, weeds, wood chips, worm castings	Use 6–30 centimetres deep. These mulches are best mixed.
Coarser mulch for large plants	bark, bracken, cardboard, carpet, cloth, feathers, fur, hair, hay, leaves, long grass, manures, paper, plant wastes, slag, stones, straw, water weeds, weeds, wood chips, wool	Use 10–30 centimetres deep.
Very coarse mulch	carpet and underfelt, concrete slabs, leather scraps, metal, stones (large)	These kill problem weeds, but most are best covered with neater mulches.
Mulches that decay quickly	bones (crushed or processed), compost, grass, grass clippings, hay, human wastes, leaf mould, manures, mushroom waste, silage, water weeds, wood ashes, worm castings	Mulches that decay quickly are good for fruit and vegetable gardens where soil fertility is vital.
Mulches that decay medium quickly	agricultural and industrial by-products, bracken, fish wastes, hair, packing materials, paper and cardboard, peat moss, pine needles, plant wastes, slag, straw, weeds	These mulches can be mixed with fast decaying material to provide a balanced mulch.
Mulches that decay slowly	bark, cloth, carpet, feathers, fur, leather scraps, leaves, plastic, sawdust, seashells, seaweed, shells and hulls, twigs and branches, wool	Use these mulches around perennials.
Mulches that decay very slowly	concrete slabs, metal, stones, wood	Concrete slabs and stones are decorative; use metal and wood where looks don't count!
Materials that may need to be buried	fish wastes, human body wastes, meat scraps	It is best to avoid these materials completely in city gardens.

1

What is Mulch?

Magnificent, magical, miraculous and mellow—mulch is all this and more. On the other hand, it is rarely malodorous, messy, mischievous or misbegotten.

Mulch is even mellifluous.

What *is* mulch?

Actually, we all know what mulch is. We've all seen mulch, walked through mulch, scuffed our feet in mulch, listened to mulch scrunching and crunching.

We just may not have recognised it.

DEFINING MULCH

According to Oxford, the authoritative British dictionary, the word has been around since the 1600s. Originally, it seems to have referred to the wet or half-rotten straw put around plants. It was used as a verb (in the sense of 'to mulch') in the 1800s and leaves and manure were referred to.

Travelling through Middle English, 'mulch' probably comes to us from a German word meaning soft, mellow or beginning to decay. Sooner or later, all mulches reach that state!

It's interesting that the word for compost comes from the French, suggesting that the Norman conquerors may have tended their tidy compost heaps while the ignorant Anglo-Saxon peasants mellowed their mulches.

My dictionaries, published in the seventies and eighties, define mulch as vegetable matter spread around plants. Webster, the standard American dictionary, mentions strawy dung. Just as the use of mulch has exploded recently, so has the range of materials that are used. Dung doesn't have to be strawy any more.

Common usage nowadays defines mulch as anything that is used to cover the soil for the purpose of suppressing weeds and promoting plant growth and/or soil fertility. Mulch can range from petrochemical spin-offs like plastic to vegetable matter like hay to animal wastes like wool and manure. Types of mulches used depend on the primary purpose they're intended to serve:

1. to improve soil fertility;
2. to suppress weeds;
3. for landscaping.

MULCH IN THE PAST

Many gardeners use some mulch, whether it's their lawn clippings, a bit of compost or the dag-ends of somebody's chook litter. I'm sure this has always been so. From the earliest days of agriculture, ten thousand to eleven thousand years ago, some mulching is likely to have been carried out, even if it was only the dumping of grain stalks on the ground at harvest time.

Mulching has been practised in Asia and Europe for centuries. Weeds have long been harvested and used in China, for example, while seaweed was commonly gathered along the coasts of the British Isles. Plant wastes were frequently left on the fields to rot and

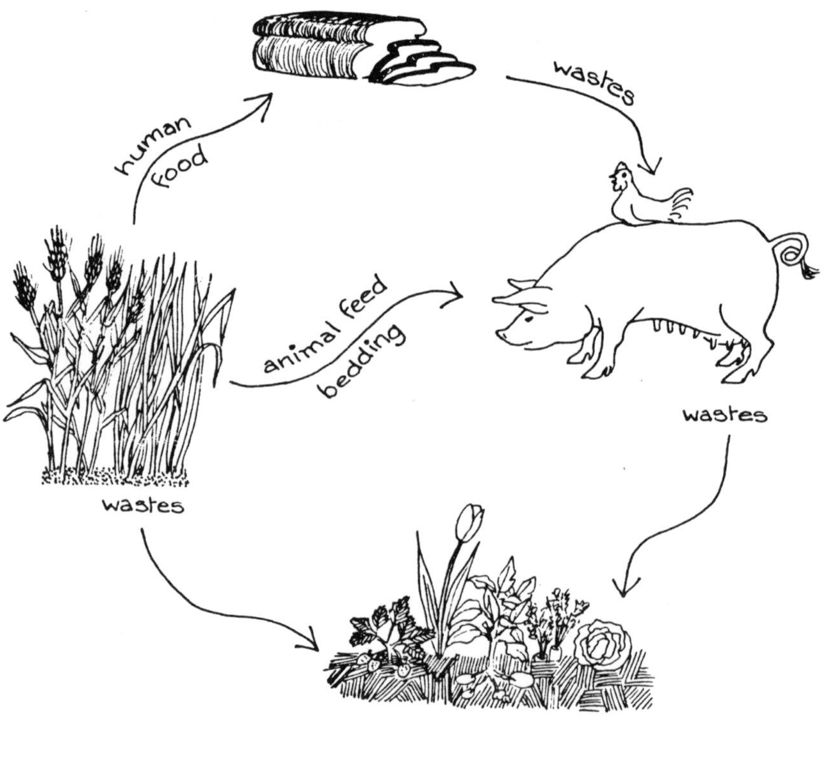

farmers everywhere understood the value of their animals' manure.

Many old-time gardeners, those in their sixties and more, tell me they've always gardened organically—long before the term 'organic' was coined by American J. I. Rodale to describe non-chemical gardening. Mulching formed part of their repertoire, along with digging materials into the soil.

Others, however, welcomed the whole range of twentieth century wonder chemicals that were said to magically fertilise their gardens and deliver death to their plants' enemies. They sent their grass clippings to the tip and carefully raked up and disposed of the autumn leaves littering their lawns. No mess for them—and the superphosphate was always there for their plants; if not super, then some special mixture for roses, vegetables or whatever.

Even those who didn't turn to chemical mixes for their plants would swear that you couldn't have a garden without digging. I still have people tell me this.

Ruth Stout began year-round mulching in Connecticut, USA, in 1942. Like a lot of caring gardeners, she had been using some mulch, but would have it ploughed under in spring, ready for an orgy of planting and sowing. One spring the guy who did the soil-turning didn't come. The mulch lay there and did what mulch does best—stopped most weeds growing and kept the soil moist and perfect for seeds and plants.

Ruth Stout got the message but she found a lot of opposition to her ideas once she began writing about them. How could you have a decent garden without turning the soil?

Ruth Stout was still gardening and mulching at ninety-four. I've seen a photograph of her looking frail but competent, surrounded by her plants. A keen gardener doesn't die, she just fades into her soil.

Esther Dean is a Sydney-sider who has also been using mulch and no-dig gardening for decades. Like Ruth Stout, she was a pioneer in modern organic gardening. Everybody talks about mulch now and 'green' ideas may be fashionable in the 1990s, but they weren't in the forties, fifties or even sixties.

These women shook up the gardening world. You can't have a garden just by throwing rubbish on the ground! Can you?

Although nominally retired, Esther Dean still travels extensively around Australia giving talks on her method of gardening. I'm reliably informed that you have to fight your way through a luxuriant jungle to get to her unit!

Permaculture One, written by Bill Mollison and David Holmgren, gave the same message, but not until 1978. By then the time was beginning to be right and permaculture didn't stop at one book. An association was set up, further books and a quarterly magazine, which is still being published, followed.

I am grateful to all these people and more. Whether or not I have read their books, whether or not I know their names, without them I would not be mulching now. My comprehensive gardening book, *The Reverse Garbage Garden,* fits into this venerable tradition of mulching books.

It is important that in promoting mulch we do not indulge in a holier-than-thou attitude. People should not be made to feel guilty or anxious about their gardening choices. It cannot be assumed that mulching is the only worthwhile form of gardening.

WHO IS USING MULCH?

When I took a look around the district I live in, I realised that a surprising number of organisations and individuals use mulch. For a start, most of the plantings carried out by the shire are mulched, so I contacted a councillor to get the lowdown on this.

She told me that she had personally started the programme of chipping the roadside branches and trees which the shire had to remove. The shire decided not to buy its own mulcher/chipper and it hires one when needed. The wood chips go onto shire gardens: outside the shire offices, beside and in the middle of streets, around public toilets and buildings like the art gallery.

My shire also puts grass clippings onto the ornamental gardens beside the toilet block.

As well as using their own chips, the shire mulches some areas with commercial chips. Like the rest of us, it can't produce all the mulch it wants.

Many local businesses use commercial mulches in the small gardens outside their premises.

Forty kilometres away in another small town, the secondary school and hospital have mulched gardens.

Most private gardens are not mulched, but some are—many more than were mulched even just five years ago.

The story seems to be the same everywhere. Commercial mulches are often used for weed control and landscaping by businesses and also around blocks of flats. Councils frequently chip some of the branches they have to remove and may well buy other mulch. Mulch is commonly used on steep roadsides to stop soil erosion and allow plants to get started.

However, mulch use is patchy throughout Australia. It is almost never utilised in broad-acre farming or even in smaller scale market gardens.

The bulk of Australians still don't think of mulching, despite the large amount of potentially useful waste material we generate. It's got better, but there's a hell of a long way to go yet.

A friend's son, who was recently in Switzerland and Zimbabwe, reports that mulch is rarely used in these countries but for very different reasons.

In Switzerland, manure and straw from animals is used by farmers and virtually all the towns process their human excrement. The resulting sewage sludge is used on pastures and forests but, because there are signs that it is contaminated with heavy metals, its use on vegetables and small fruits is restricted. Apart from this, there seem to be neglible amounts of organic matter available for mulching so, despite an ethos of waste use, there just isn't enough mulch around to be noticeable.

He reported that in Zimbabwe he saw huge mounds of maize stalks piled up and unused. There was no sign at all of mulching there.

I spoke with a field officer of an aid organisation, Community Aid Abroad. She was just back from the Horn of Africa. She said that the people there gardened without any kind of fertiliser at all, with the obvious result that the soil fertility has dropped and dropped. Because of drought and war, they grow no vegetables or fruit, contenting themselves with a simple grain crop. The stalks are then fed to livestock or left to lie on the ground, but there is not enough over to act as a mulch or fertiliser. Animal manure is used for fires, although there are also few animals—almost no chooks, for example. No cabbage leaves for the compost bin or cut grass for mulch!

Basically the farmers are too poor, as well as too ignorant of soil conservation, to give anything back to the soil.

Mini-dams, emergency food supplies, seeds and tools, an improved standard of living and education about soil were her recommendations.

Bob writes from Tasmania that 'although we have a mulch ethic here, it's not usually carried through. People complain about their soil but never do much about it.'

A lot of talk, but not always much being done. That about sums up what I've found out, too!

2

Why Use Mulch?

There are several excellent reasons for using mulch, at least to some extent, in your garden.

1. to improve the soil;
2. to reduce the need to water;
3. to reduce the work load;
4. to suppress weed growth;
5. for landscaping;
6. because it is a good use of many waste materials.

WHY I MULCH

Before I go on to discuss my six reasons for mulching, I'll explain how I came to mulch my own garden.

As an idealistic twenty-year-old I moved to the country and started my first ever garden. I already loved raw fruit and vegetables and determined to grow my own. Naturally I gardened organically—I never thought of doing anything else. I am lucky in never having had to switch over from chemicals to organics. I threw myself into reading about organic gardening and did my poor best. Like all gardens, mine had its failures and, I suppose, a few successes. Enough for me to keep going.

I was shocked to find that the farmers' wives living around me used poisons on their gardens. I am not a fanatic: I accept there are good times, places and reasons to resort to poisons—but dieldrin on your roses?

I still can't understand it.

If a market gardener uses superphosphate and poisons, well—it's his living, not a hobby. If Africans, devastated by war, drought and poor farming, need chemical fertilisers to get crops going—so be it. I only hope that education and growing understanding will make people look at alternatives, think sidewards a bit.

But dieldrin on roses!

Since then I've had many gardens in many different places, learning all the time. I started using bits of mulch. The bits became more—combined with blood and bone and a little compost—and this became the fertiliser I relied on. Kitchen wastes, animal manure and our own weeds all went onto the garden. Many gardeners stop here.

At my present home I have no water for my garden. The climate's certainly wet—over 100 centimetres a year—but it nearly all falls in winter, turning everything to quagmire and mud. I used to get a good spring garden going, struggle with weeds, give everything up over the dry summer months, then wait the wet winter through.

Of course, mulch helped to keep the weeds in check in my garden beds, but they *would* keep invading from the pathways. One day it struck me that I should mulch the pathways—a revelation! I've never looked back. That was about ten years ago and I've haven't bought a potato at the shops since. When friends moan about the price of potatoes, I don't know what they're talking about.

Since my paths are permanent, they never have to grow plants. Consequently, I can use mulches on them that are good for suppressing weeds, but not high in nutrients. I mulched my paths with paper, cardboard and sawdust, all of which I got free. Of course, I couldn't generate that much mulch myself, so I quickly learnt how to collect other people's throwaways.

I am now quite used to looking out for grass clippings, leaves, newspapers, sawdust and seaweed wherever I go.

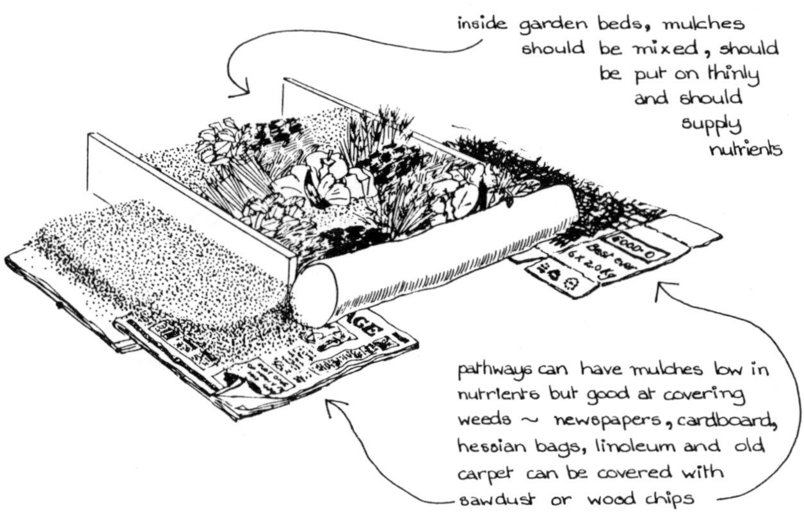

Not only did the year-round mulch solve my weed problems, it also helped most plants to grow well in summer. Not as well as if they were watered, I admit, but well enough to keep me in beans, peppers, tomatoes, zucchinis and the marvellous cabbage family. In winter the mulch stops soil loss due to heavy rain and eliminates mud. Summer or winter, worms love it under mulch.

I have been describing my berry and vegetable garden. I also mulch fruit trees as well as I can manage (never as much as I'd like) and ornamental plants are mulched during their first year to help them get going.

Control of weeds and year-round growing are only the most obvious benefits of mulching. I also work less and the work is easier—carting grass clippings around, rather than digging and weeding. No more backache! And my garden houses many creatures, from blue-tongue lizards to toads and worms. Most magical of all, the mulch supplies everything my plants need to grow.

Everything.

So, I'd be a fool not to mulch, wouldn't I?

REASONS FOR USING MULCH

1. MULCH IMPROVES THE SOIL

Mulch helps the soil in a multitude of ways. Firstly, it keeps it moist. The mulch forms a physical barrier between the soil and the drying sun and wind. Just slip your hand under mulch on a hot day and test this for yourself.

Similarly, it stops the driving rain compacting the soil or washing it away.

Quite obviously, mulch adds organic matter, decayed and decaying, to the soil, not only in the narrow interface area where soil and mulch meet, but also deep down. Worms drag organic matter down into the soil and this improves the tilth of the soil. Subterranean activity creates channels for air and water.

Water-soluble nutrients slowly leach out of the mulch into the soil and the root zones of plants. These nutrients vary enormously for the different kinds of mulch. Newspaper and sawdust, for example, are low in nutrients, while grass, manure and weeds are high. Nutrients continue to be made available as the mulch decays.

All these factors combine to encourage activity in the soil. If you feel around under a heavy mulch, you should be able to find worms — the most noticeable form of soil life. There are also grubs and larvae (not all of which you will welcome!) and a teeming invisible population of bacteria, fungi and the like. Love 'em or hate 'em, these creatures are essential to a healthy soil. Their activities make many nutrients available to plant roots and they also add their own invaluable body wastes.

I believe that blue-tongue lizards spring fully-formed from mulch, as do frogs and toads, insects, insect-devouring birds, skinks and probably many other creatures I don't notice or can't name. Mulch provides haven, shelter and, indirectly, food for above-ground life. Slugs may be encouraged by mulch, but so, too, are the birds that enjoy eating slugs.

Lastly, you may have heard, perhaps with trepidation, about soil pH (see also page 36). pH measures the relative acidity or alkalinity of soil: 7.0 is neutral, lower is acid, higher is alkaline. Many plants like a neutral soil (most vegetables, for example, prefer a slightly acid to neutral soil — 6.5 to 7.0), some like it acid, some like it alkaline. It sounds like a minefield, doesn't it?

Either too acid or too alkaline is destructive for soils and plants. Luckily, organic matter in the soil slowly corrects pH imbalance,

but also, and even more importantly, it protects plants against imbalances so they can perform well when the pH is not ideal.

Mulch painlessly delivers that organic matter.

To recap, mulch improves the soil:

1. by keeping it moist;
2. by stopping damage from heavy rains;
3. by adding organic matter;
4. by adding nutrients;
5. by encouraging all forms of under-soil life;
6. by encouraging birds, insects, frogs, lizards etc.;
7. by slowly correcting pH and helping plants to cope with less-than-perfect soil.

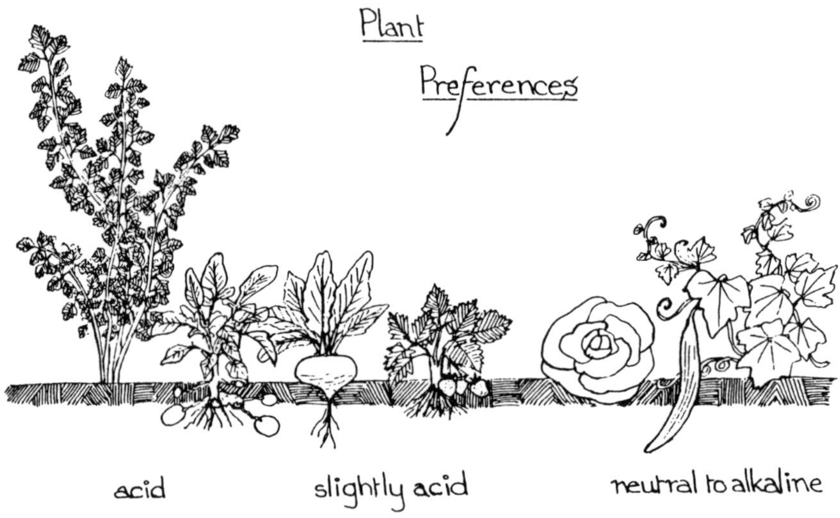

2. MULCH REDUCES THE NEED TO WATER

Because mulch covers the ground, it prevents evaporation of moisture from the soil by both sun and wind. It also gives *your* plants sole rights to what moisture there is by preventing weeds from competing.

Well-mulched, a surprising number of plants will survive weeks of no rain and months of low rain. In my edible garden this includes: asparagus, beans, brambleberries, cabbage family members (broccoli, brussels sprouts, cabbage, cauliflower and Chinese cabbage), carrots, Chinese gooseberries, eggplant, garlic, herbs, marrows, onions, passionfruit, pepino, peppers, potatoes, pumpkins, raspberries, rhubarb, strawberries and tomatoes.

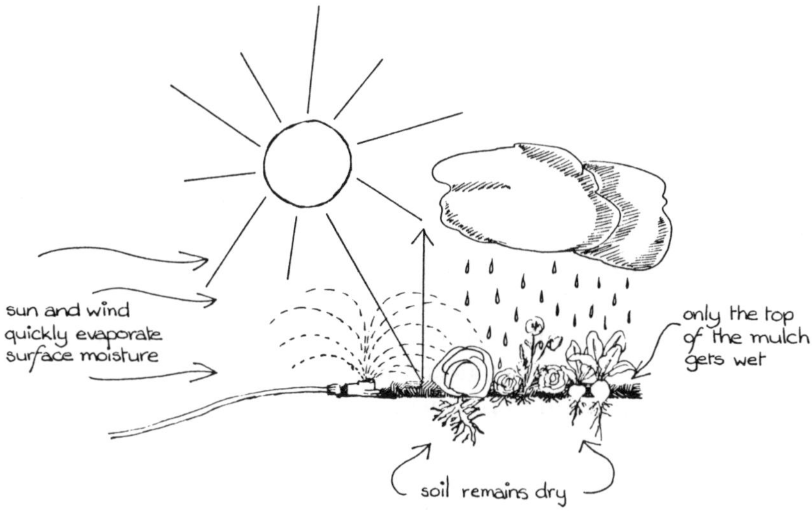

Mulching reduces the need to water, even if you have the ability to do so. This saves water, which may well save you money, and also saves you time. City gardeners tell me less water equals less money, and anyone can see that less water equals less time spent messing around watering.

Water is a commodity that is often in short supply in Australian cities. In drought periods, Melbourne, the park capital, languishes under water restrictions. Lawns fade and vegetables wilt. Water problems aren't solved, they're just deferred by wetter summers. Water is only going to go up in price.

With mulch, sprinkling is not likely to be an effective way of watering. It will cool the plants, which may be the intention, but most of the water sits on top of the mulch, not penetrating into the soil, and quickly evaporates off. This can also happen with very

light rain. However, anything lost by light rain not penetrating the mulch is offset by decent rains being retained better.

Water should be delivered to where it's needed. Individual plants can be given a good soaking by leaving a hose running into the mulch beside them. Alternatively, containers with holes can be sunk into the ground beside plants and these can be kept filled—the water then seeps out slowly. Soaker hoses can be set permanently under or over the mulch and water leaked out slowly, or microjets can be used.

However the water is delivered, less is needed when you mulch. If you don't or can't water at all, mulch will enable most of your plants to survive, even to prosper.

3. MULCH REDUCES YOUR WORK LOAD

The most daunting task gardeners face is weeding. Life comes to resemble an unremitting battle, a fight to the death against these hardy garden invaders—invaders and, all too often, conquerors.

Mulching is not only one of the most effective ways of defeating the enemy; it is also the easiest. Ninety per cent (this is a guess, not a scientifically tested and proven fact) of weeds can be completely controlled by mulch. Just cover them heavily and keep them covered.

Some mulches bring in their own seeds: grass, hay, manure and weeds are examples. However, seeds will not germinate or, should they germinate, will not be able to grow, so long as you keep the mulch topped up. Seeds and seedlings must be continually covered and smothered.

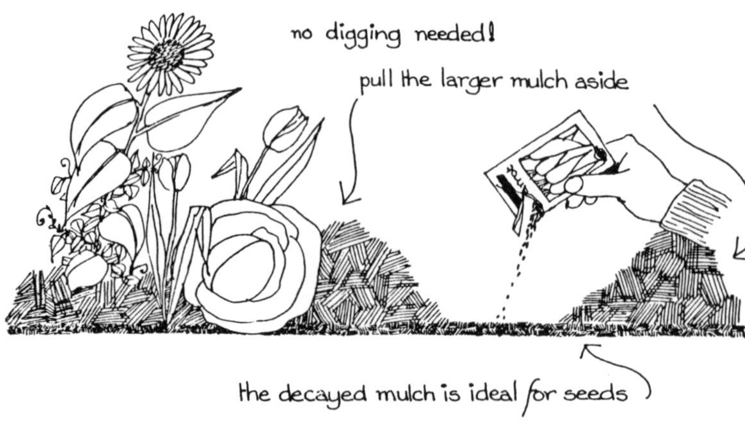

Plenty of mulches, like hulls, lawn clippings, sawdust, seaweed and commercial mulches, are seed-free.

Seeds remain in the soil, frequently viable for many years. If the soil is cultivated or dug, the adult weeds are cut up and killed, but seeds are brought to the surface where they germinate with a vengeance. I never have a worse weed problem than when I dig a garden bed. With mulch you no longer have to dig and this means that weed seeds are doomed to languish underground.

Talking of digging, this also is hard work which can be largely avoided by mulching. Digging incorporates air into your soil, but mulching effects a slow improvement in soil fertility and tilth which makes digging unnecessary. Soil that is teeming with life forms and well supplied with decaying organic matter doesn't lack oxygen.

Perhaps you want to dig in order to create a fine seedbed? Unnecessary. If you pull mulch aside from a bed that has been continuously mulched for six months, you will find underneath a well-rotted layer that is perfect for seeds.

The labour of digging and weeding are replaced by collecting, transporting and spreading mulch. If you haven't used mulch, try it. You will quickly discover how much less work it takes and how much easier that work is.

4. MULCH SUPPRESSES WEEDS

This is indisputable.

A heavy covering of mulch smothers growing weeds. The only exception to this are some particularly pushy types that creep along under the ground, such as couch grass and sorrel (see page 56). Weeds like this can be a problem.

So—mulch isn't perfect, but it almost is.

Not all mulches have the same covering power, just as not all mulches supply loads of nutrients. Grass clippings can hardly be expected to smother blackberries. However, discarded carpet will. This is why it is usual to combine different kinds of mulch. Weeds might find their way through wood chips, but not if the ground is first covered with plastic. Paper, cardboard, cloth and hessian bags work well but look awful—unless covered with more attractive mulches.

Many mulches, like sawdust and wood chips, also make it hard for wind-blown seeds to germinate. Of course, if you continually apply mulch, these seeds get covered, anyway.

5. MULCH IS GOOD FOR LANDSCAPING

Some people think that mulch must always be messy and smelly and covered with flies. Perhaps they forget that stones and wood chips, when laid on top of the ground, are as much mulch as kitchen wastes and manure.

Mulches are used in landscaping for two main reasons: to suppress weeds; and to look good.

Mulch is attractive.

This can be just as true of the mulch you put in your vegetable garden as of the commercial bark and wood chips used by businesses, car parks, councils, hospitals, schools and supermarkets. It doesn't matter what you put on the soil, people only see the top layer. Plastic is ugly, but not when covered by 20 centimetres of chips. Kitchen waste and manure can be covered by grass clippings, hay or straw for a thoroughly attractive finish. Paper doesn't blow around if it is covered by other mulches.

More attention should be paid to the effects created by different mulches. The range of colours, shapes and textures is enormous. Straw, for example, gives a clean, bright, rustic look. Pebbles create a calm, austere atmosphere. The colour of a mulch can be used to highlight, blend in with or offset the colours of buildings, fixtures and plants.

Seed pods and bark chips can be a rich, varied red; wood chips come in cool greys (just right to blend in with the Australian plants they came from); sawdust is a light reflective yellow; seaweed is dark and nut shells are lively mixed colours.

Of course, mulches tend to fade with time. So do paints.

Mulches are also used in places where there are no plants—under play areas in parks and schools, for example. Here the mulch is

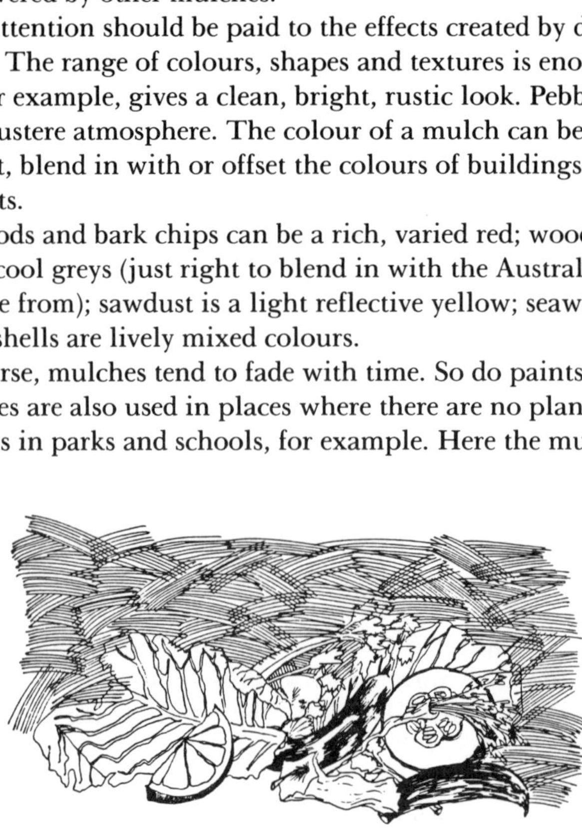

kitchen rubbish can be covered with any tidy kind of mulch

used as a weed suppressant, but also because it is a safe surface for kids to play and fall on. Naturally, stones aren't chosen for this kind of mulch, but soft materials are laid directly on the ground, or put over plastic.

There is certainly a mulch for every situation. However, there is probably more than one that will do any job, so it is worthwhile looking around and keeping an open mind. And watch out for the range of colours and textures.

6. MULCH IS AN EXCELLENT USE FOR MANY WASTE MATERIALS

On television the other day I heard that Tokyo Bay is filling up with rubbish. Residents have to separate rubbish into the recyclable and unrecyclable, and put it into *clear* plastic bags so everyone can check that they've really sorted it. The authorities even wanted to make them put their names on the bags so that the neighbours could shame those who were careless or lazy!

In other parts of Japan, citizens must sort their rubbish into six different piles! The categories are something like: aluminium, glass, organic rubbish, paper and card, plastics and other.

There have also been strong European initiatives on recycling. This is not surprising since in Europe there are many industrialised urban communities living high on the hog. They *make* a lot of rubbish. Germany has been seen as a leader in recycling.

Most Australians have some access to recycling collections. In 1991, eighty-six per cent of Victorian households had weekly kerbside collections and ninety-six per cent of Sydney households had such access. In other states and territories, there are drop-off bins, centres or depots. It's reassuring to believe that our copious waste is being re-used.

But is it?

It makes good economic sense to recycle aluminium—it's cheaper to recycle than to make from raw materials. Not so for most other waste. An enormous amount ends up as landfill.

Paper makes up about forty per cent of rubbish in American households and about twenty per cent in City of Melbourne households. Collection rates are probably quite high but what happens after collection? There is a world-wide glut of paper. Most of it can't be used. The high costs of collection and transportation make recycled paper products uncompetitive compared with paper from raw materials. Limited demand for the recycled end-product means limited demand for waste paper. Consumers seem unwilling to pay recycling costs via the product price so it's into landfill with a lot of collected paper and it appears paper doesn't even rot in well-run landfill.

Germany's recycling industry is in turmoil because the citizens have thrown themselves into it with German efficiency. So much waste is collected that it can't be used. There is an economic problem here that has to be tackled. It costs to recycle. Who's paying?

Only a few councils are beginning to try to recycle organic wastes, yet food and garden rubbish amounts to forty-five per cent of the total in City of Melbourne households while sixty-five per cent of the rubbish consists of garden and kitchen wastes plus paper. Even if you carefully sort your rubbish and put it in the right place at the right time, you can be sure that little is recycled.

Why not re-use it yourself and reduce your household rubbish by sixty-five per cent? Garden wastes, kitchen wastes and paper/cardboard—these are the very easiest wastes for *you* to use.

Aluminium, glass and plastics must rightly be left to the professionals but anyone can learn how to mulch a garden with their own wastes. You can't tell other people what to do with their

garbage, but you *can* ask friends and neighbours if *you* can use some of it. If they're concerned about the environment they'll be pleased to help you and *you* will be creatively using *their* rubbish.

If you cut grass, rake up leaves or trim twigs and branches, you can put this garden waste to good use, too.

Even indoor plants benefit from light mulches such as coffee grounds, grass clippings and leaves—*everyone* can manage to re-use at least some of their rubbish and many gardeners can collect and utilise rubbish around them.

If the appearance of your garden is vital, just make sure that the messier material is hidden under tidy mulches.

Who knows—*your* example could set your whole street sorting and using their rubbish!

There are incontrovertible reasons for mulching. It improves your soil, reduces the water needs of plants, reduces your work load, suppresses weeds, is useful in landscaping and is a creative method of using the most common items of household rubbish. And . . . if you're as stout of heart as Ruth, you may find mulching allows you still to be gardening in your nineties!

3

Where to Get Mulch?

Where to get mulch from depends on the kind you're looking for. There are three categories of mulch:

1. commercial mulches;
2. free mulches;
3. bought materials which are not necessarily sold as mulch.

Naturally, these three types of mulch are found and collected in different ways. There are times, places and reasons to use each of them. Commercial mulches are frequently used by businesses and organisations and for landscaping. They are neat and tidy, available in large or small quantities and can often be delivered. Commercial mulches can be obtained when you want them in the quantity you need. They normally decay slowly, have good weed-covering value and look attractive.

I prefer to use free mulches. They are very suitable for vegetable and fruit growing and can be mixed and matched with different plants and situations. In using free mulches, which are nothing other than somebody's rubbish, you are making use of waste materials commonly thrown away. Moreover you can easily make sure that you include mulches which contain plenty of the nutrients needed by growing plants.

Free mulches are collected as and when they are available, not necessarily when you need them.

The third category overlaps with the second. Sawdust can be had free from carpenters, joiners, some sawmills, schools and woodworkers. It can also be bought from sawmills in truck loads delivered to your property. It is frequently bought by farmers who use it around dairies and animal pens, so it is not only sold as a mulch.

Hay is the same. It is most often bought to feed animals, but a gardener can make good use of it. Although most gardeners I know pick up old carpet for nothing, sometimes it is available at a low enough price to make it worth buying for garden use. So this category includes any material that is bought for a variety of uses, but which can also be used as mulch.

COMMERCIAL MULCHES

These are the easiest to get hold of. They are available in amounts ranging from a small plastic bagful through trailer loads right up to truck loads.

Small, pre-packaged lots can be bought at plant nurseries, hardware shops and even many supermarkets. If the cost is off-putting, you can buy a small amount every week until you have achieved the mulch cover you want. However, this is certainly the most expensive way to buy commercial mulches.

It is cheaper to buy these materials in bulk. Contractors, garden shops and plant nurseries are the places to contact. Many suppliers deliver trailer loads or offer free use of their own trailers if you live within a specified radius of the business. This is an excellent option if you wish to establish a smallish garden quickly. The whole area can then be mulched straightaway and planted out at your convenience.

Delivered truck loads offer the lowest price per square metre and the largest quantity. However, you must check that you need this amount and also make sure the truck has access. Ask the contractor or driver how much space he needs. Delivery may be too difficult in the inner city but, in the suburbs and the country, it could be the best way to landscape in a hurry.

Get the contractor to explain how much material there is in a truck load and what area of garden it will mulch adequately. There's no harm in overbuying (the material will keep), but it costs of course.

Commercial mulches are available any time you want them, just by paying your money over.

FREE MULCHES

In *The Reverse Garbage Garden*, I was solely concerned with free mulches—how you get them and how you use them. I divided the areas from which you can gather them into three zones. I haven't had any better ideas since then.

1. around your home and garden;
2. within your neighbourhood and from friends;
3. in your local area, including wherever you go for recreation and work.

I discuss different mulches in detail in Chapter 5, so I will just mention the more common ones here.

Zone 1

First, look around your house and yard. Some wastes can be found in *all* households: kitchen scraps and paper, for example. Keep looking and you may find cardboard, discarded clothing, garden waste, grass clippings, human hair, stones, weeds and wood ashes. If you have a woodpile for feeding a fire or stove, there will be bark, chips and sawdust. If you do woodwork, collect the shavings as well as the sawdust.

Animals? You may have anything from stud cattle to bantam hens, from pigs to pigeons. Whatever you have, there will be manure and possibly bedding, which is excellent because it incorporates urine (except in the case of birds, since they don't urinate).

Do you have fence lines where leaves and fallen grass accumulate? Do you have bushes and trees which drop leaves and twigs?

Most of us have mulching material all around us. It is just a matter of teaching ourselves to look for it.

If you can't use all your material at once, store it. Paper, for example, is difficult to use unless you have something to cover it with. Not only is it messy, but it blows around.

Get to the point where all the sixty-five per cent of your household rubbish which consists of kitchen and garden wastes and paper is being used on your garden. This may be all you need if your garden is tiny or your mulching aims are meagre. However, when you find how your garden thrives on mulch, how the soil improves, the worms multiply, the bushes flower and the trees fruit—and all this for so little work!—you may want more.

Easy. You now move into zone 2.

Zone 2

Zone 2 begins with the same sort of rubbish, but this time you relieve your friends and neighbours of their embarrassing bounty.

How?

Ask them. Environmentally-conscious friends who somehow

Gardens in front of our shire offices are kept weed-free with wood chip mulch.

This small, inner city backyard is easily mulched and stays neat and tidy.

Mixed mulches protect this vegetable garden from both drought and heavy rain AND supply nutrients.

Worms flourish under mulch.

A blue-tongue lizard is hiding in the garden mulch.

Frogs love the cool environment of a mulched garden.

never get round to using their rubbish themselves will love you for doing it for them.

I'll add here that it is fatal to imply criticism of others. Don't make anyone feel guilty. God knows, we're all just living our lives as best we can. The last thing we need is a so-called friend making us feel like environmental vandals. Just be thankful that *you* benefit.

It is difficult to use somebody else's kitchen waste unless you can collect it regularly and put it on the garden promptly to avoid maggots and messes. I don't bother. However, I do have friends who save me their newspapers. This is very easy.

Cut grass, garden trimmings, grass clippings, leaves, sawdust, weeds and wood ashes can all be sought. When your neighbours find you are taking the grass clippings they usually dump behind their back fence, they will almost certainly *feel grateful* to you.

It's true. There you are selfishly collecting *their* fine mulch for *your* garden and *they* thank you for it! It makes them feel better about dumping it. If you ask for clippings from people you know or have seen mowing their lawns, odds are they will thank you for saving them a trip to the tip. At the very least, it will mean they don't have to put it out on the right day for garbage collection. Many people offer to bag clippings for you.

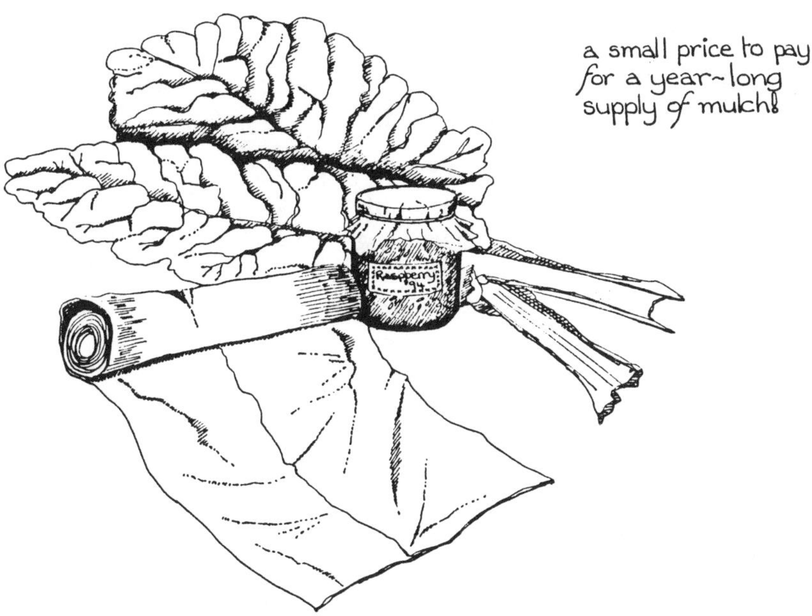

This is the archetypical win/win situation and really does make everyone happy. Offering to supply garbage bags is a good idea as is the occasional jar of jam or bunch of silver beet.

Suggest to friends that you clean out their animal pens for a share of the bedding and manure.

Nobody knows instinctively that you are looking for mulch. Don't forget to ask.

You can find that you have more mulch than you can use immediately. I try not to turn any away, so I often end up storing hay and paper (see pages 100–1). Try not to store inflammable material against your house where it will be a fire hazard. In the country, material like this can harbour snakes, too, so find a safe spot for it and protect it from the rain.

Now you may well have all the mulch you need. I don't know why you have to be so undemanding!

I want more.

Zone 3

Thus, out of the Twilight Zone and into—Zone 3. It's when you are working in Zone 3 that sober citizens may say of you: 'Who was that masked scavenger I saw?'

Well, not quite, perhaps.

Zone 3 includes the rest of the world, or what you can reach of it. It covers beaches, businesses, industrial areas, parks, pine plantations, relatives' places, roadsides, shopping centres and wasteland.

Chapter 5 has a list of many of the mulches you can get free. In a shop window in Fitzroy, very much an inner city suburb of Melbourne, I saw a sign which read:

MOVING HOUSE?
Suitable
Packing
BOXES
available here
for
FREE

The shopkeeper doesn't care whether or not you are moving house, she just wants to get rid of the cardboard boxes. Chemists, hardware shops, supermarkets—most shops, in fact, have cardboard boxes coming out their ears. The local fruit shop that sells my raspberries can re-use some of their boxes. The rest go into a *huge* container and end up at the tip. Also thrown out are the solid brown paper bags that potatoes come in—these cover weeds well and also rot quickly.

Talking of fruit shops, they junk awe-inspiring amounts of less-than-perfect fruit and vegetables *all the time.* I collect some of this rubbish for feeding guinea pigs and rabbits, but it can also be used for mulching the garden. Make sure that, if you say you will collect rubbish regularly, you do so. Otherwise, shopkeepers will get annoyed.

You won't notice this kind of rubbish as a customer. It's normally lurking somewhere out the back, so ask about it. Markets have laneways full of it, rivers of vegetable waste—check them out.

Lateral thinking is called for. Ask yourself what wastes do businesses and individuals create that you can use?

If it's waste, the odds are they'll be pleased to have you take it away.

Try flower shops for plant wastes, hairdressers for hair clippings, poodle parlours for dog hair, shops for organic packing materials like shredded paper, carpenters and joiners for sawdust and wood shavings. Dressmakers may have scraps of material and your local council may make grass clippings, leaves and wood chips available—if you ask.

When you approach people, remember they'll think you're crazy. Explain what a keen gardener you are and murmur about mulch. They'll *know* then that you're crazy, but harmless!

Never seem to criticise them for creating waste. They won't like it.

Farms offer the possibility of animal bedding, grass, manure, plant wastes, spoiled (or poor and weedy) hay, straw and weeds. Check if spraying programmes could have affected the material you're getting. Horse stables, pig farms, poultry sheds and all forms of factory farming can provide bedding and manure.

Visit factories and businesses around you. Peer into their backyards. Keep your eyes open.

Grass is often cut on roadsides and in parks and wasteland and left lying in lumps. Feel noble when you take it, as piles of cut grass stop the new grass from regrowing underneath them. I feel free to take grass and leaves from public land. If they're on private land, even if they are obviously unwanted, I find someone to ask.

Pine needles are not needed by plantation owners but are excellent in parts of your garden.

Finding mulch is not a matter of luck or the star sign you were born under. It's an attitude. If you decide to get free mulches, you will. They are there—everywhere; in the inner city, in the suburbs, in the country, on flat land and hilly, inland or the coast. Where *you* are is where the mulch is. Just look for it.

Free mulches are unlikely to be delivered to your doorstep. You will have to collect them (see pages 97–100 for more about collecting mulch). I have a city friend who brings home leaves from a park in plastic shopping bags. This is fine for a small garden. Ashes, grass, leaves, paper, sawdust—all can be collected in small quantities.

Slightly larger quantities can be transported in shopping trolleys, prams or pushers—anything on wheels. The car boot is good and always with you when you drive somewhere, whether it's for a day at the beach or a visit to the fruit shop. It's inadvisable to put material straight into your boot. I do it, but my husband is threatening to leave me and even I don't think the boot is an appropriate place for grass to grow in! Use bags of some sort. Strong plastic bags such as garbage bags or recycled animal feed and briquette bags, either hessian or synthetic, are good.

For larger amounts, trailers and the backs of utilities come in handy if you have them. However, I mulch a large garden completely with free mulches and only use my car. It *is* a station wagon, which helps.

With free mulches, you are unlikely to get huge quantities all at once. The drip-feed method of mulching is more practical—small amounts brought in frequently.

BOUGHT MATERIALS, NOT NECESSARILY SOLD FOR MULCHING

This is an area where creativity is called for as these materials may not be marketed specifically to gardeners. You have to recognise their potential and seek them out.

Since, like commercial mulches, they are bought, they can normally be delivered. If you want sawdust, for example, sawmills will deliver truck loads. They also sell smaller quantities, such as trailer loads, which you can pick up yourself.

Sawmills are obviously one place from which to get this kind of material. Contractors and farmers are another. They have hay, spoiled hay and straw. Mushroom farms have mushroom compost; worm farms have worm castings. Contractors who remove and chip shrubs and trees sell the bark and wood chips, while stables supply horse manure. Contractors can deliver pebbles and stones.

The best way to find out about materials like these is to look at what other gardeners use. See also Chapter 6.

MIXING MULCHES

There is, of course, no need to use any category of mulch exclusively. You can buy second-rate hay from a contractor to spread under your bushes and fruit trees, tuck kitchen rubbish under a neighbour's grass clippings on your vegetables, and have a trailer load of bark chips delivered for the rose garden out the front.

Don't let anybody fool you into thinking that there are inflexible rules you should follow.

You might use paper under sawdust in one spot, yet decide on plastic under bark chips in another.

Whether you find your mulch lying around the roadsides, lurking in a truck or neatly pre-packed, there's always a place, time and reason for slapping it (somewhere) on your garden.

4

How to Use Mulch

It is reassuring or disturbing, depending on your point of view, that everything can so often be reduced to numbers. There are three categories of mulch, three zones for finding free mulch and now there are two basic ways of using mulch:

1. as one part of your total garden-attack plan;
2. as *the* method of gardening.

MULCH AS ONE COMPONENT OF YOUR GARDENING STRATEGY

This is the easiest way to use mulch, and the way most of us started. Mulch is used as and when we have it, and is more or less slapped onto the garden. We are unlikely to be actively seeking mulch from friends and neighbours, but we grab it if it falls into our laps.

If this is the way you garden, make sure you utilise all your own waste. Remember that sixty-five per cent of your household rubbish is likely to be made up of garden and kitchen waste and paper—all easily incorporated into your garden.

Learn how to use all the different sorts of mulch that may come your way. Every bit you use decreases your need for other forms of fertiliser.

At this stage you will combine mulch with weeding and perhaps cultivation or digging, and with the use of fertilisers. These can be either organic or non-organic. Most of the gardeners I know who mulch have come to it through organic gardening. However, there is no reason for non-organic gardeners not to mulch. It can serve as a transition to organics and, hopefully, will at least decrease the need for artificial fertilisers.

Combining non-organic gardening and mulch is complicated when you also want to spray for pests. Mulch encourages a diversity of insect and animal life to flourish in the garden. Pesticides kill them. Consequently, the chemicals undo a lot of the good the mulch accomplishes.

If you are a sprayer learning to use mulch, try to hold off on the chemicals. Improvement won't happen overnight and patience and alternative strategies are called for. Go on using superphosphate if you must, but try hand picking pests, avoiding problem plants and finding out how timing affects pest infestation. I cover the question of organic pest control in *The Reverse Garbage Garden*, pages 112–16.

Mulch doesn't get rid of your pests, but it does encourage a thriving little ecosystem to get going which incorporates their predators as well as the pests. This, don't forget, is an artificial

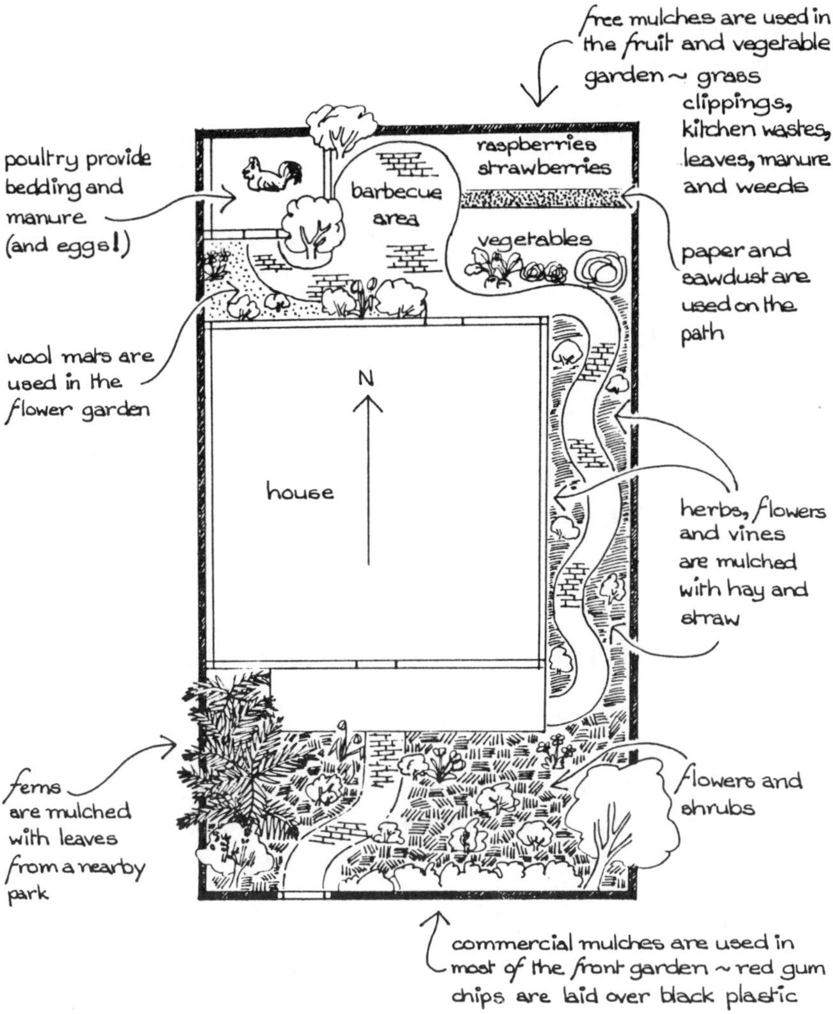

ecosystem which relies on you. You have to keep the mulch coming, the cats out, rotate your plants, remove diseased specimens and watch for what works and what doesn't.

Organic gardeners combine mulching with digging, weeding and the use of fertilisers like blood and bone, compost and commercial organic preparations. Mulch is a useful tool for the organic gardener. It helps keep down the weeding needed, makes the soil easier to dig and reduces the need for other fertilisers.

Another halfway house with mulch is to use it in only part of the garden. Mulch used for landscaping can consist of just plastic and commercial wood chips in the front garden, which is seen by the whole neighbourhood, while the backyard is left to the weeds and nature. At the opposite extreme, many gardeners mulch their fruit and vegetable areas with high-nutrient mulches like manure, but leave the ornamentals to get by on whatever the soil offers.

Neither of these approaches is ideal. If you accept that mulching benefits your soil and garden, you should aim to use it in all areas eventually. On the other hand, we are none of us perfect and must make whatever compromises we find practical. I have an order of priorities. Once the first is mulched, I mulch the second on the list and so on. I don't always get to the bottom of the list. My mulching order goes:

1. vegetables;
2. small fruits;
3. fruit trees;
4. anything else.

Some people decide to run with the mulch. It's not difficult to make the jump from being a sometime mulch-user to a fanatic.

MULCH AS A TOTAL GARDENING CONCEPT

This is the move to year-round, garden-wide mulching. Mulch can do virtually all the work of promoting plant growth, soil improvement, water conservation and weed control. It is the way to go if you want to get the maximum advantages from mulch and if you are prepared to cultivate an attitude of always, but *always* looking out for mulch.

I know of two ways gardeners practise this.

1. plants are grown *through* heavy mulch;
2. plants are grown *on top* of heavy mulch.

Planting through mulch

This is how my small fruit and vegetables are grown.

In the vegetable garden the whole area is mulched, paths and all.

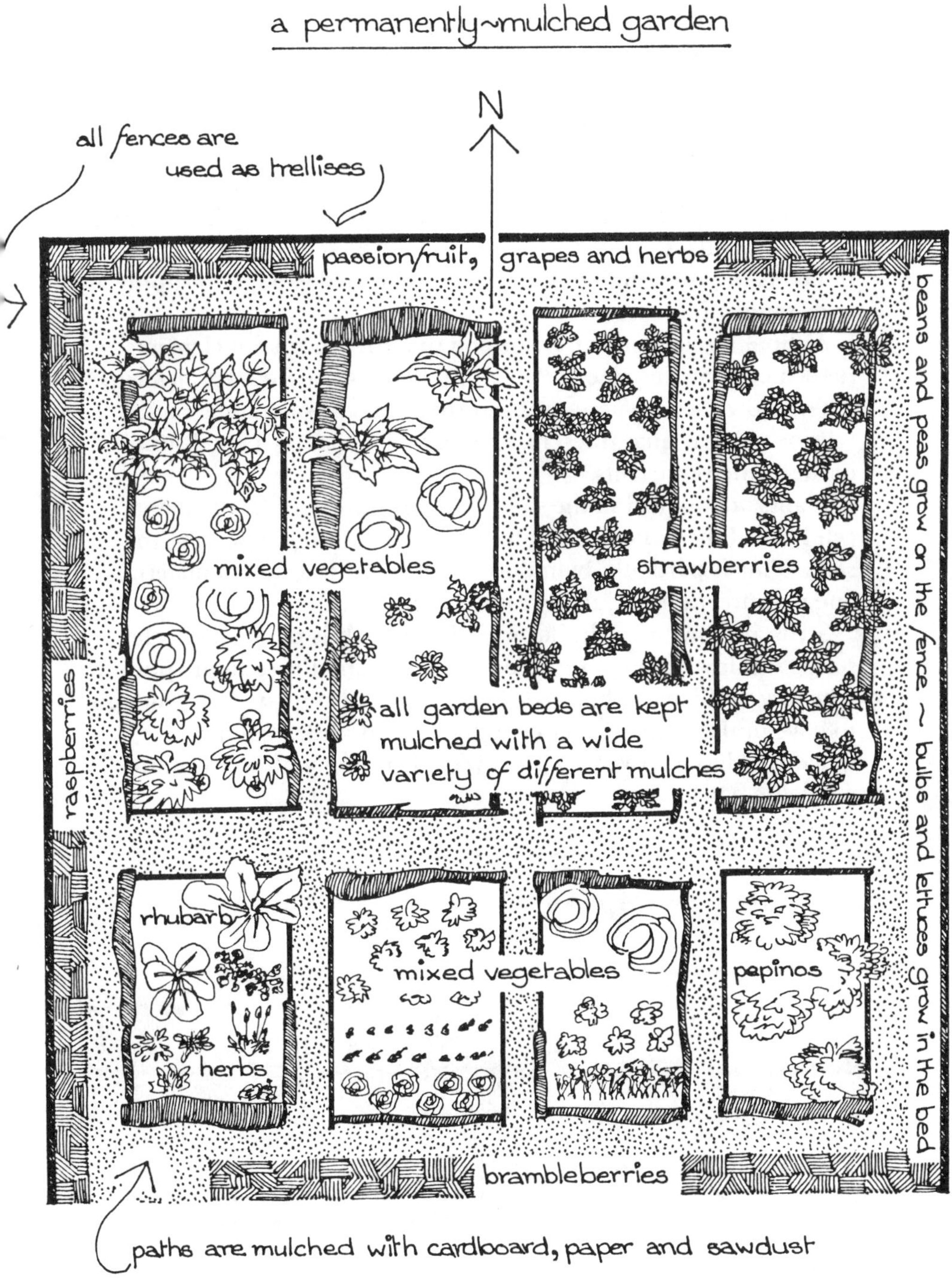

The beds are permanent, never alternating with the paths, and this allows the decayed mulch to build up into a raised bed that is never compacted by being walked or wheelbarrowed over.

On the paths I use low-nutrient mulches, mainly cardboard and paper covered with sawdust. These are weed-free, totally destroy the weeds under them and make it difficult for wind-blown seeds to germinate. Other gardeners I know of use wood chips or, in Queensland, blady grass.

Mulch on the paths has to be replaced every year or so.

All kinds of mulch are used in the vegetable beds: compost, feathers, grass, hair, hay, leaves, manure, pine needles, plant wastes, seaweed, shells, straw, weeds and wood ashes. It's important that the mulch is heavy enough to kill weeds and varied enough to supply a wide range of nutrients.

Although scientists can name and quantify the major elements needed by growing plants, they can't do the same for trace elements. They know they are necessary, but they don't always know how, why or in what quantities. Mixing mulches ensures a good supply of these micronutrients.

Plants are planted in the soil. Mulch is pulled aside, the plant put in, and the mulch pulled back close to, but not touching, the stem. This is particularly important with moist or rotten materials and manure, as they quickly rot the stem.

Seeds are sown in a similar way. The mulch is pulled back, the soil is shallowly dug or scraped to soften it and the seeds are put in. The row will have to be weeded until the seedlings are large enough to have mulch pulled up close to them. Neil of Briagolong (page 125) has discovered a good way of coping with weeds in his vegetable patch.

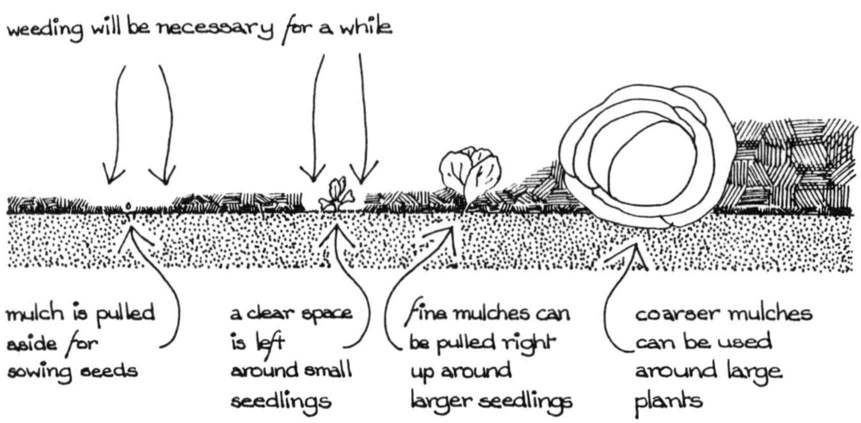

See also *The Reverse Garbage Garden* pages 15 and 73–9.

After the beds have been continuously mulched for several months, it becomes much easier. When you pull the mulch aside, there will be no weeds growing and the soil will be covered with finely-decayed mulch, ideal for sowing seeds or putting plants in.

I work in the same way with the rest of my garden, but rarely get around to mulching the paths there. I mulch along my beds and plant bulbs, bushes, flowers, shrubs and trees by pulling the mulch aside in chosen spots.

When landscaped gardens are planted by businesses, councils and organisations, they normally put plastic down first and cover it with bark or wood chips. These can still be unrotted a decade later but they are low in nutrients. However, combined with plastic, they do a good job stopping weeds growing. Plants are put into the soil through holes or slits in the plastic.

This is not a method I recommend for growing fruit and vegetables, which need constant nourishment and soil improvement. Plastic doesn't contribute anything apart from weed control and perhaps moisture retention so the plastic method is best for tough plants that can 'get by' well.

The essence of the 'planting through mulch' method is that thick and constantly replenished mulch is laid on the ground. This is pulled aside or a cut or slit is made in it and plants are planted directly into the ground.

Planting on top of mulch

The alternative method involves putting a thick layer of relatively impenetrable mulch on the ground first. This is often cardboard,

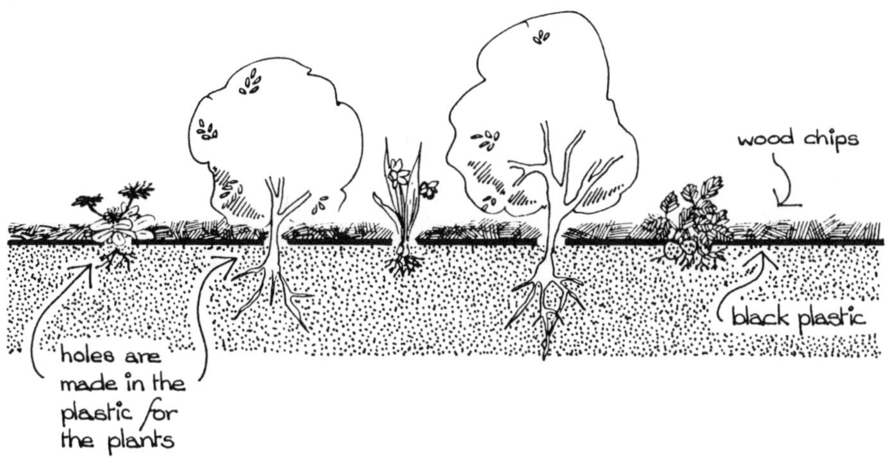

cloth or paper. Nutrient-rich mulch is then spread across the first layer to some depth—20 centimetres or so. Plants and seeds are put in the layer of good mulch and grow above the impenetrable layer. Keith and Maree of Kirwan (page 121) use mulch in this way.

The sustaining mulch on top consists of material such as compost, grass, hay, leaf mould and leaves, manure, mushroom waste, seaweed, straw and worm castings. Seeds can simply fall through the loose mulch and get buried but this is avoided by placing a handful of soil where the seed is to be sown, or a row of soil for a row of seeds. The soil is only deep enough to put the seed in. It is the mulch that nourishes the growing plants.

It is probably a good idea to sprinkle blood and bone on or into this mulch.

You have to keep on mulching your garden or weeds will begin to grow in the rich mulch layer. If you want to stake plants, the stakes should be driven right through all the mulch into the ground.

Gardeners often use this method for starting a new organic garden since it means they don't have to worry about deficiencies in the soil or problems with drainage.

Although perennials *can* be planted and the mulch including the lower layers will eventually rot, letting the plant roots forage into the soil, I can't think of a good reason for not poking a hole in the mulch and putting perennials into the soil itself. If a reader wants to write and berate me about this, feel free.

As well as richly supplying nutrients to the plants, this method keeps their roots up above the ground so that drainage is not a

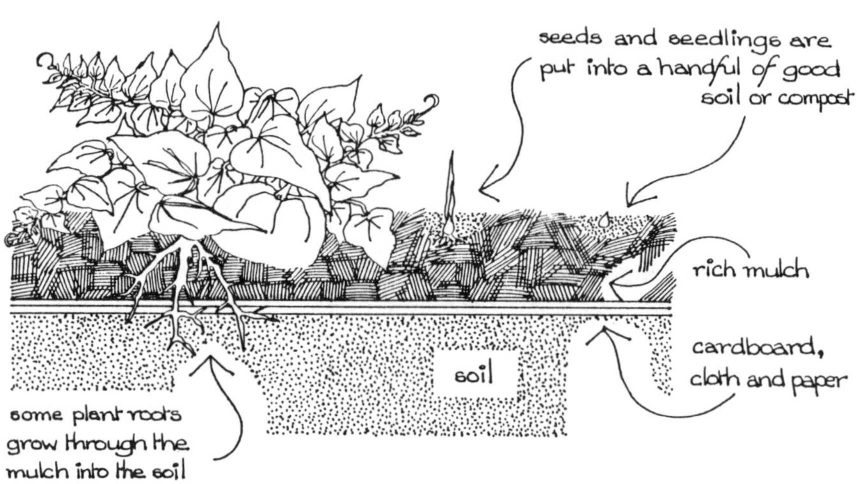

problem. In any case, raised beds of any kind are easier to garden with than those level with the ground—there's not so far to bend!

I believed this sort of mulching to be a new discovery, until I watched a film called 'Man of Aran', which documents and dramatises a centuries-old lifestyle still in existence when the film was made in the 1930s. Aran is rocky with little soil and the people are shown as living on fish and potatoes. To grow their potatoes, they collected seaweed and laid it on the bare cracked rock. They then dug soil out of cracks and crevices, where it was washed by rain, and sprinkled it lightly on the seaweed base. The seed potatoes went into this mixture.

My problem with this method is lack of water. I have no summer water at all and need to encourage my plants to wriggle their roots as deeply into the soil as possible to protect them from the dry spells. Where roots have the choice of soft succulent mulch or of digging through paper to get to the lacking-in-lots-of-things soil, where are they going to go? Too right—into the mulch. They will then tend to dry out without water.

The other problem is that you need to get a lot of material together before you can start. However, you can design your garden to be created in sections as mulch becomes available.

You can't use many of the commercial mulches for this sort of garden as they tend to be low in nutrients and to decay very slowly. You need mulches that will rot quickly in order to release the elements your plants must have. Bought mulches would be more likely to come from category 3—materials like spoiled hay and straw, plus of course manure from farms and stables.

This is a terrific way to get a highly productive garden going in a hurry.

MULCH SECRETS

These are the secret ingredients gardeners withhold to ensure *their* garden is better than yours. The secret is that there aren't any!

Mulch is dead easy to use.

One of the terrific things about mulch is that it is hard to go very wrong. Using chemical fertilisers seems much more difficult to me. Then you're dealing with mainly highly-soluble nutrients—if your plants don't OD you could end up fertilising the town water supply.

Can't do much harm with mulch.

Experience is certainly the best teacher, but I can give a few pointers I have gleaned from my own.

Varying your mulches

There's not much that can go wrong with mulch and the few problems I've encountered can be avoided by mixing the mulches and by putting mulches on thinly.

Some mulches mat down, becoming a solid interwoven barrier that stops rain getting through and that decays slowly—rather like felt, which is made from fibres matted together rather than knitted or woven. Grass clippings and pine needles can do this.

The solution is simple: never put heavy layers of mulch on your garden. Constant thin layers keep the weeds down just as well, but don't mat.

Another possible problem is lack of adequate nutrients for plant growth. This is solved by using a variety of mulches. If they are applied in thin, mixed and matched layers over your garden, the whole garden will get a wide range of nutrients. If you cover one bed with half a metre of horse manure and another with half a metre of sawdust, the first has more nitrogen than it needs while the second gets none.

Spread everything thinly.

There is no need to be obsessive about what mulch goes where. You don't have to keep maps of where you last put the manure or the seaweed. Just keep spreading it around.

Plants that are lacking in nitrogen are stunted and the leaves look yellowish. If you suspect this is the case, sprinkle blood and bone around or water the plants weekly with manure tea (see *The Reverse Garbage Garden*, pages 68–70)—this provides an instant boost. Put your next lot of high-nitrogen mulch in this bed.

Nitrogen versus carbon

Mulches vary enormously in their nutrient value. They can be roughly divided into those with high nitrogen and those with high carbon content. Ideally (remembering we live in a far from ideal world) these should be mixed. Manures, which are all high in nitrogen, are best covered by a high carbon mulch such as hay. This stops them smelling, releasing nitrogen wastefully into the air and attracting flies (and your dog!).

High carbon mulches add bulk and organic matter to your soil while high nitrogen mulches tend to release more nutrients faster for your plants. You can see why both are good.

Animal bedding is an example of a mulch combining both. The hay or sawdust used in animal pens is soaked in urine and brings the manure with it as well. Telecom Australia has been doing some experiments using shredded telephone books as animal bedding.

This later ends up as a fertiliser. An excellent idea!

High nitrogen mulches include: alfalfa hay, blood and bone, compost, feathers, fish waste, fur, grass clippings, green weeds, hair, legume wastes, manures, nut shells, sewage, urine and wool.

High carbon mulches include: paper, sawdust and wood.

In the middle somewhere are materials like: cloth, fruit wastes, hay, leaves, plant stalks and wastes and straw.

Remember that all these mulches are useful. Many country gardeners are fixated on manure. They have an obsessive belief that they *must* get manure for their garden, especially for vegetables. Manure is certainly useful, but it is only one among many.

Generally speaking, high-nitrogen materials tend to decay more rapidly than low-nitrogen ones, but this is not always the case. Nut shells, for example, couldn't be slower.

Paper, sawdust and wood chips are best used when water conservation and weed suppression are the major aims. Vegetables, in particular, grow furiously and crop heavily. They need mulches that rot quickly and supply nutrients.

I have updated the quick mulch reference from *The Reverse Garbage Garden* and included it here (pages x–xi) so that you can see at a glance what the various mulches are good for.

Mulches that sustain or remain

Some mulches rot quickly, some slowly, some very slowly. Mulches that decay very slowly are suitable for permanent plantings, such as around shrubs and trees, while mulches that rot quickly are advisable for annual beds.

Stones will be very much in the way if you mulch your carrots with them.

If you want landscaping and weed control, choose mulches that decay slowly. Commercial ones can be good value and their cost can be spread over a decade or so.

Mulches that decay very slowly include: bark and wood chips, concrete slabs, many nut shells, stones, uncrushed bones and sea shells.

Mulches that decay relatively slowly include: bracken, cloth, feathers, fibrous kitchen and plant wastes, fur, hair, hulls, leaves (some), paper, pine needles, seaweed and wool.

Mulches that decay rapidly include: bones (crushed), compost, grass clippings, hay, some kitchen and plant wastes, leaf mould, manures, mushroom waste, weeds and worm castings.

Decide what you want from your mulch and choose accordingly. In many situations a mixture is best.

Mulch and pH

pH is the measure of the relative alkalinity or acidity of your soil (see page 9). With most soils, it isn't necessary to know the pH so long as you look after the soil by mixed and continuous mulching.

Old-time gardeners used to speak of soil being sweet or sour and they used to taste the soil to determine which it was. In the same way lemons, being acid, taste sour.

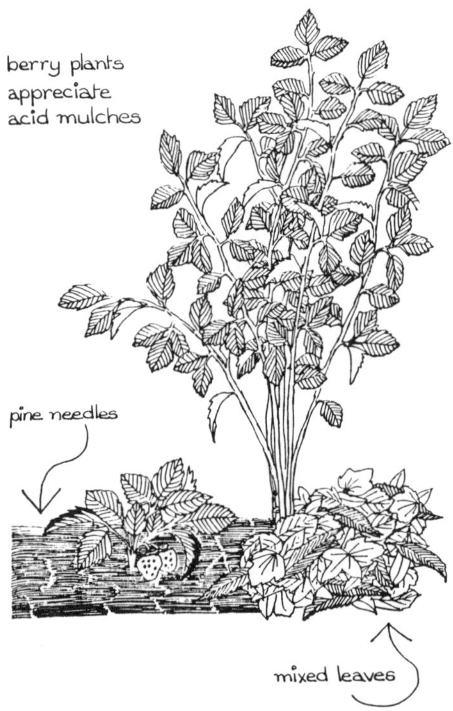

A few mulches are acid. You can either correct this or use it to your advantage. Blueberries, for example, demand an acid soil. Mulch them with pine needles and you should have no trouble with this rather demanding plant. All berries like a slightly acid soil, so pine needles and leaves can be used on them.

If pine needles or oak leaves are used for plants with a leaning towards neutrality, like most vegetables, sprinkle them with limestone or wood ashes.

Wood ashes are an excellent source of potash for your garden. There is no harm in sprinkling them from time to time over all areas, except those growing acid-loving plants. They are highly alkaline so never put them down thickly or in a lump.

Mulch and stems

Don't let mulch touch the stems of shrubs and trees or the stems may rot. Leave about 20 centimetres bare all around. Weed this small area.

Don't bring mulch up close to small seedlings for the same reason and also because they are easily killed if the mulch covers them. As my vegetable seedlings grow, say by the time they're 20 centimetres high, I pull mulch right up around them, but not hot, moist or rotten materials or manure. Clean grass, hay or straw is good.

When not to mulch

Mulch keeps your soil moist and cool. This is great in summer and fine over winter when you're growing cold-tolerant plants. In fact, I'd hate to go through winter here without mulch, because the soil would be battered and washed away by heavy rains.

In spring, you may want the soil to warm up quickly so you can plant summer annuals. Simply remove all the mulch from the beds you want warmed. Within a few days to a week, you will be able to feel the difference in the soil temperature. Gradually replace the mulch as your plants or seeds grow.

I never mulch my seedbed. I create the seedbed, using good soil from my garden beds, and I change the soil after taking several lots of seedlings from it. A fine mulch would be okay, but since I don't want mulch to smother my seeds, and I have to water the seedbed regularly anyway, mulch serves little purpose.

Different mulches for different uses

If you cover fine seedlings with 50 centimetres of hay you are unlikely to be happy with the result.

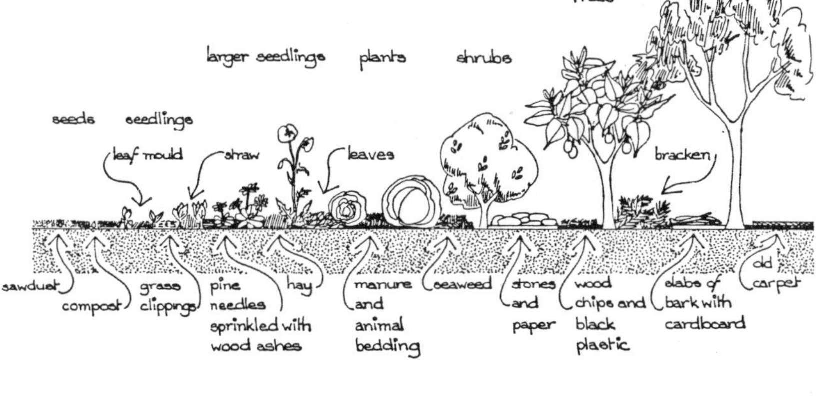

Fine mulches should be used with seeds, coarser but still fine mulches for small seedlings, and very coarse mulches are best for shrubs and trees. Compost is terrific for those seeds that would have been smothered by the hay, but try spreading it around your orchard and it's not likely to go far enough to do much good.

Mulches suitable for covering seeds: compost, leaf mould, peat moss, sawdust and worm castings.

Mulches suitable for small seedlings: compost, grass clippings, leaf mould, mushroom waste, peat moss, sawdust and worm castings.

Mulches for larger seedlings and plants: bark (shredded), hair (small quantities), hay, hulls, kitchen wastes, leaf mould and leaves, long grass, manures, mushroom waste, nut shells, packing materials (small or shredded), peat moss, pine needles, plant wastes, seaweed, straw, weeds, wood chips and worm castings.

Mulches for shrubs and trees: bark, bracken, cardboard, carpet, cloth, feathers, fur, hair, leaves, long grass, manures, paper, plant wastes, stones, straw, weeds, wood chips and wool.

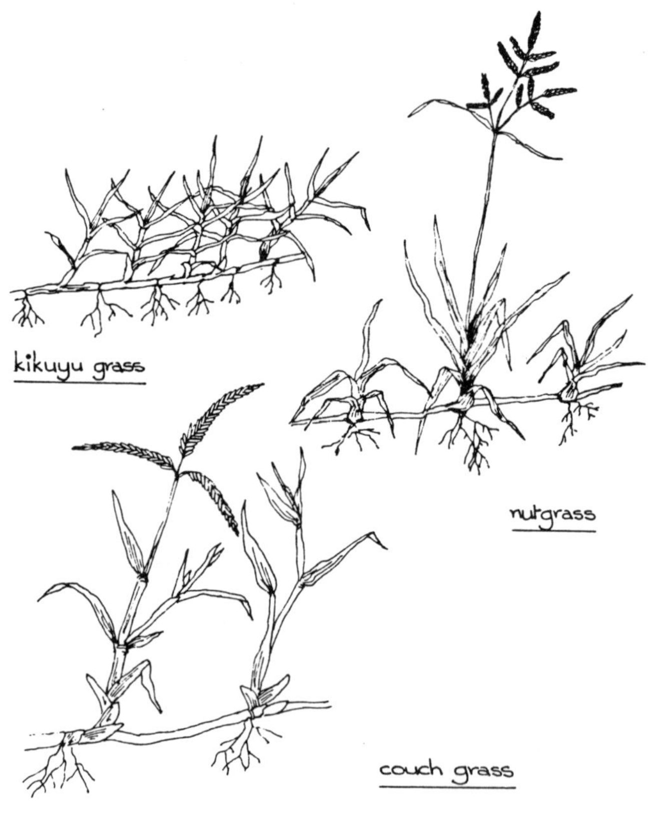

Weed alert

I have mentioned using weeds as mulch. Nearly all the weeds that are pulled can be put back on the garden as mulch. The exceptions are weeds that regrow from stem and root pieces. If you use these, you may create a *big* weed problem (see page 56).

Look out for plants like couch grass, nutgrass, onion weed, oxalis, sorrel and wandering jew.

Do not use bits of these plants as mulch and be prepared to battle them if they are already in your mulched garden. Heavy mulching is only partially effective against them. A wide (at least 1 metre) and well-maintained mulch barrier can be used to stop them getting in. Use heavy-covering materials like cardboard, cloth and paper—put the mulch on thickly and keep it topped up.

If they're in your beds already, dig the weeds up as well as keeping the garden mulched. Try to get all the underground pieces out.

Lynn from Brisbane provides some useful information about her battle with nutgrass (pages 121–2).

Taxed with the failure of mulch to eliminate such weeds, Ruth Stout answered something to the tune of: It doesn't wash your dishes or make your beds either. So what?

Weeds that are normally considered problems because of their hefty roots, like dandelion and dock, are stopped dead by heavy mulch. If thrown on as mulch, they will certainly regrow from root pieces, but they don't spread and are easily removed. The whole plant can be used or the roots discarded and the leaves (which are rich in minerals) used.

Mulch achieves so much and saves us so much work, that its inability to eliminate nutgrass and sorrel from our lives is a minor drawback.

5

Five Hundred Free Mulches!

Well, maybe not *quite* five hundred, although I *wish* I knew that many.

Most people don't realise how much mulch there is around.

A friend was astounded when I told him he could put grass clippings on his garden. He simply hadn't thought of it. Another gardener believed that grass clippings were bad because they mat down. I told him the rule about putting on thin layers of mulch frequently.

These two people were keen to mulch and both decided to start using their grass clippings. They only needed someone to tell them they could.

People believe things like: Mulch makes soil acid. It stops rain getting through. You can't use paper. Pine needles are bad because, after all, nothing much grows in pine forests. (See Chapter 9 for the answers to these and many more questions about mulch.)

We don't all have access to everything and we won't all choose to use the same mulches. I can't name every mulch there is, nor do I have all the answers: not even all the questions. However, I'd like to help gardeners look around with open eyes.

Many of the mulches mentioned below are listed in *The Reverse Garbage Garden* which gives details about using free mulches in your fruit and vegetable gardens, with specific references to all the major types grown. Here the mulches are organised differently because this book is exclusively about *mulch* and this chapter presents *the most important information* for all you keen mulchers.

Mulches have been, sometimes rather arbitrarily, divided into those that decay quickly, those that decay slowly (and very, very slowly) and those that sit in the middle. Often mulches straddle categories. With nut shells, for example, walnut shells decay slowly while peanut shells decay quickly. Sometimes I have simply thrown a dart at my compost pile to choose which category to put a mulch in!

FREE AND NEARLY FREE MULCHES THAT DECAY QUICKLY

Bones

Bones provide their living owners with a kind of reservoir of minerals that are stored up in the good times and can be called upon by the body in bad times. They are rich in magnesium, calcium and phosphorous. They contain from two to four per cent nitrogen which is released slowly to plants.

Most of us eat enough meat for the bones to make a decent mineral contribution to our gardens. We can also get them from friends, farmers and local abattoirs and butchers. Even vegetarians can use bones obtained from outside sources.

Bones put on your garden in an unprocessed form look odd, decay slowly and would certainly attract carnivores like the dog next door. They are best crushed and sprinkled on garden beds. An ordinary household meat grinder can be used to crush small quantities. Dry them in the sun for a day or two to make grinding easier.

Bones can also be dissolved. Alternate 20 centimetre layers of wood ashes with layers of bones in a forty-four gallon drum or a plastic rubbish bin and cover the whole with water. Leave this until the bones are soft, then mash them into the water and spread the resulting paste on the garden. (I doubt if it will ever replace Vegemite as a sandwich spread!)

Cover this material with other mulches.

Commercial blood and bone contains burnt, pulverised bones and is good for seeds and seedlings.

Compost

Compost decays quickly as a mulch because it is already decomposed or partly decomposed. It consists of many of the mulches listed here but they have been piled together, whether scientifically or not, and encouraged to decay. It is rich in nutrients and trace elements which vary with the materials used in its making, the speed and nature of the composting and even the season of the year.

Mulching is a way of turning your whole garden into one gigantic compost bin!

In its simplest form, compost can be made by simply piling (mixed) wastes together in a heap on the ground. Grass, kitchen wastes, manure and weeds are all good ingredients. Blood, blood and bone, herbal teas (see *The Reverse Garbage Garden*, pages 68–70), urine and wood ashes can be sprinkled through the heap as it is built up. Compost can be used partly decomposed or left until it resembles rich soil.

You can get fancy and use bins made of almost anything (even hay!) or buy commercial bins.

Compost is terrific for covering seeds, mulching around tiny seedlings and as a booster for newly-transplanted seedlings. It is not particularly good at covering weeds but is very high in nutrients. If you plant on top of your mulch (see pages 31–3), seeds and seedlings can be put into a handful of well-decomposed compost.

A rich garden tea can be made by putting a handful of compost in a bucket of water for a week. Water your plants with it to give them a boost. For more information about making compost see *The Reverse Garbage Garden*, pages 58–68.

Grass and hay

Long grass and hay are coarser than clippings and lower in nitrogen. They don't decay super quickly like compost but do seem to disappear pretty smartly. They are a medium coarse mulch and are good around half-grown vegetables or bushes and trees. They can be applied very thickly and provide a mini windbreak when pulled up close to plants.

Grass and hay bring seeds into the garden, especially hay. Keep putting mulch on top so that the seeds can't germinate.

Long grass often grows around trees, along fences, in waste ground and building sites. I suggest in Chapter 8 how to find, gather and transport this sort of material.

Spoiled hay can sometimes be had for the asking from farmers. This is hay that has got wet and can't be fed to stock. Otherwise you can buy spoiled and poor, weedy hay cheaply from contractors, farmers and produce shops. Check the current local price to make sure you're not paying too much. It's easy to get stung. Spoiled hay is still okay if it's got grass growing on it! Just make sure you're paying rock bottom prices.

When you are putting slabs of hay on the garden, shake the seeds out first, then cover them with the slabs. If slabs of hay that are spread on your garden start sprouting a crop of weeds, simply turn them upside down, either by hand or using a fork.

Hay provides a lot of organic matter and worms love it. Legume hay is the highest in nitrogen. Hay is neat looking and cheerful. It quickly loses its straw colour and goes a pleasant, earthy brown. Long grass dries to a bleached colour.

Grass clippings

These are rich in nitrogen and contain potash. They rot quickly—keep applying them in thin layers to stop them packing down. Alternatively, they can be mixed with a coarser material like bark chips or sawdust and applied in thick layers.

On their own, they do a better job of suppling nutrients than of killing weeds, but they can be put on top of cardboard, paper or plastic for better weed control.

Being quite fine, grass clippings can be used close to seeds and small seedlings. However, they're good anywhere in your garden—around ornamentals, berry plants and even under fruit trees if you have enough. Your roses will like them.

Grass clippings are my mainstay mulch. I rely on them. Neil from Briagolong (see page 125) is known as a crazy organic gardener in his village. Everyone drops grass clippings on the nature strip outside his house. He gets so many he has huge compost piles as well as masses for mulching. Sometimes he has so much he has to put the word out for Briag residents to lay off for a bit! Not that he ain't grateful!

I happen to know where a half-dozen or so households dump grass in my own home town—and I'm not about to tell anybody else!

Human body wastes

Sewage sludge is a useful by-product of treating large-scale human waste. It's widely available in Europe and America and can now be obtained in some cities in Australia. It is often used as a top dressing on pasture and forests. As it often suffers from heavy metal contamination its use on home gardens is limited.

In Hobart sewage sludge is mixed with plant prunings to make a useful garden product, while in Canberra sewage sludge is dried and then burned at high temperatures to sterilise it. Sewage sludge and ash (such as Canberra makes) can be used, with caution, in your garden.

What about your own household waste? Compost toilets that have been approved by councils are available in Australia. They produce clean humus that can be used on gardens no matter where you live. The material is like compost. It is fine and rich in nutrients. Put it around seeds and seedlings. You are unlikely to have a lot of it.

Although fresh urine is virtually free from bacteria, human faeces is teeming with the things. This is not to say that human faeces cannot be used in the garden, but that it must be used with great care unless it has first been properly composted in a compost that reaches 70 to 80 degrees Celsius, or otherwise treated.

I spent five years burying the family's excretions under perennials or plants that would stay put for a year or two. Don't plant root crops in these beds for several years. Human faeces should never be buried in areas that drain into water supplies and don't leave it lying around on the surface of the soil. If nothing else, your dog (or the neighbour's) is sure to roll in it.

Many people have told me that they shovel out their septic tanks, rather than hire a contractor, and wheelbarrow the stuff to their gardens. This may not be allowed but who's to know? The material doesn't stink (much!) and is at least partly digested. Don't use it in city gardens and don't spread it on vegetable beds. It is best if you cover it with other materials less rich in nitrogen, like hay and straw. An interesting bit of trivia is the fact that tomato seedlings invariably seem to sprout from it.

Urine is high in nitrogen and potash, and is renowned for killing grass and giving citrus trees a boost. Don't overapply (this is the only reason it kills patches of grass). Use it directly on mulch, in compost or around trees, but not right on the plants themselves. Diluted, it can be used to water in seedlings.

Although clean when first excreted, urine provides an ideal medium for bacterial growth. Use it quickly. A good method is to fill a bucket half full of sawdust and try to get the family to use it. As the sawdust gets sodden, put more in. Don't let it sit around more than a few days before putting it on your plants.

Leaf mould

Leaf mould is fast decaying because it is, by definition, partly decayed already; it is like a compost made solely from leaves. It is high in nutrients and especially trace elements although not so high in nitrogen. A fairly fine and dense mulch, leaf mould can be used around small or large seedlings.

Leaves collect under shrubs and trees and often against barriers like fences. Over months and even years, the leaves settle, break up and start to rot, becoming leaf mould. It's also easy to make your own leaf mould. Pile leaves into a container with an open bottom, such as a circle of chicken wire held in place by tomato stakes set into the ground. The pile must be half a metre or so high, so that the leaves are pressed down as they would be on the forest floor. They may take years to decompose. You can speed up the process by first shredding the leaves and/or by adding a high-nitrogen material like grass clippings, manure or urine.

Sprinkle leaf mould with lime or wood ashes to correct acidity, unless it is to be used around acid-loving plants such as blueberries and strawberries.

Manures

Manure comes from an old French word, 'mainoverer', related to manoeuvre. The meaning slid from 'to hold or manage land', to 'to till or cultivate' to 'enriching land with fertiliser'. From the beginning, 'manure' has been connected with fertilising the land. The Anglo-Saxon equivalent, 'dung', a typically earthy and monosyllabic word, seems to have lost the race.

All animal manures are useful mulches and, if fresh, composted or stored under cover (rather than left lying around in the open), provide many nutrients. They all break down quite rapidly, even when applied with bedding like hay, sawdust and straw.

The nutrient value of manure varies. Animals fed concentrates or legume hay produce richer manure. Young, growing animals, pregnant females, milk-producers and working animals excrete manure richer in nutrients than idle adults.

Animal bedding is generally a better nutrient source than pure manure, since urine is soaked up by the bedding. Birds like chooks, or course, don't excrete urine. Bedding also supplies bulky organic matter.

Because it is so rich in nitrogen and salts, manure is best used with high-carbon mulches. On a continuum, manure stands at the opposite end from commercial bark and wood chips. They are low in nutrients and decay slowly; manure is high in nutrients and decays quickly.

If it's not mixed with bedding, cover manure in the garden with mulches like grass, hay or sawdust to prevent the loss of ammonia, to stop the manure smelling, keep flies away and prevent the dog rolling in it!

Sawdust is an excellent free kitty litter, being highly absorbent, and the resultant mixture can be spread around your ornamentals and trees and covered with an attractive mulch.

If the manure has been gathered from paddocks, it will have lost much of its nitrogen but it will still contribute organic matter that decomposes rapidly.

I have used many different manures fresh on the garden, despite advice not to. These include cow, duck, goat, hen, horse and rabbit manure. Fresh manure is often regarded suspiciously by gardeners since it may damage or 'burn' plants because of its high nitrogen content. If manures are applied on top of the ground, along with other mulches, they do not get the chance to burn tender roots.

Manures are referred to as cold, meaning those that are relatively low in nitrogen like cow manure, and hot, meaning those high in nitrogen like pigeon, poultry and rabbit manure. Hot manures should be used more sparingly than cold and *must not touch plants*. Pigeon manure, which you can get from pigeon fanciers and racers, is so rich it should be very thinly spread or put into compost.

In many city gardens, dogs can be a real problem. If they insist on using your garden as their toilet, the best thing to do is to bury the manure (not where you grow root crops) and try to fence the dogs out. Cats are harder to keep out and like to relieve themselves in soft soil. They normally avoid mulched gardens unless the mulch is sawdust. Mulches like grass, hay, seaweed and straw will help protect your garden.

Large quantities of manure can be had from farmers, horse breeders and fanciers, pony-riding farms and stables, neighbours and friends. Offer to clean out a neighbour's chook house and you'll probably make a friend for life!

If you have access to manure from commercial poultry farms, remember that most modern hens are raised in appalling conditions and fed on growth stimulants. I would be wary about using enormous quantities. This warning applies to all feedlot animals.

Some people have expressed concerns about horse manure and tetanus so I asked my family doctor for advice. Tetanus is caused by a bacteria whose spores lie dormant in the soil. The spores can be present in any manure or soil. Contact with horse manure is no more likely to cause a problem than contact with soil.

Manure and animal bedding are rich in weed seeds. Cover them with further layers of mulch to prevent a weed problem.

Manure is not the only source of nitrogen. The continuous use of varied mulches such as compost, grass clippings, hulls, kitchen wastes, shells and weeds gradually makes enough nitrogen available even for demanding plants. However, improvement can be slow in a very neglected bit of ground. If a serious nitrogen deficiency is suspected (see page 117), and manure is not available, use diluted urine, blood and bone or manure tea.

Mushroom waste

This is another mulch which decays rapidly because it is already decomposed or partly so. Mushroom growers use a rich compost of hay and manure as a growing medium and often sell this after harvesting their crops of commercial mushrooms.

It is a good soil conditioner and can be used as a fine mulch over seeds and around tiny seedlings. As with compost, you can't expect mushroom waste to smother weeds. Combine it with mulches like grass, hay, paper and straw.

Using spent mushroom compost can even give you a bonus crop of mushrooms in your garden!

If you have access to plenty of hay and horse manure, you can make this compost yourself, either for growing mushrooms or for applying to the garden.

Silage

I have often made silage by accident when I've collected grass clippings and forgotten to put them on the garden.

Silage is normally something farmers make in pits from green plant matter—they use it as animal feed. I prefer just to put grass and weeds straight on the garden as mulch, but silage is an alternative you can try. The end product is a heavy lumpy mulch you can dump on weeds. Don't let it touch plant stems and don't try to use it around seeds or tiny seedlings.

Put a heap of grass clippings or green weeds or both on the ground, about 30 centimetres wide and the same high. Wet the pile and let it settle. You can chop it into pieces and use it when the surface has dried out.

Flies can be a problem so you will have to cover the heap with plastic. Black plastic tends to make it stink; try clear.

Water weeds

Several gardeners have told me they pull weeds out of their dams and drag them onto the garden. They use the most primitive of methods—on with the gumboots and into the water! Water hyacinth is an example of this sort of plant. Pulling some out may not solve the problems it causes by clogging up waterways, but it can't hurt.

Water plants provide moisture and organic matter, and are high in nutrients, many of which have been snatched from the water they grew in.

They are best mixed with other mulches for weed control and are useful rather than attractive. Use water weeds on annuals or perennials but not around seeds and tiny seedlings.

Reeds are another water weed, but rooted in the soil. They can be harvested by cutting rather than dragging the whole plant out. Rushes, like reeds, grow along water or in wet land and can also be cut.

You're unlikely to go looking for this sort of material, but if you or your friends have weed-clogged bits of water, do yourself a favour and clean them up.

Wood ashes

Wood ashes are composed of 50 to 70 per cent lime which is why they are so good for correcting soil acidity. They also contain calcium, phosphorous, potash and trace elements. Wood ashes vary from 6 per cent potassium if taken from the main trunk of a conifer to 60 per cent from apple tree branches. Hardwoods supply ashes with a higher nutrient value than softwoods, but this is a free material so use whatever you can get.

Because they are so alkaline, wood ashes should be sprinkled thinly in garden beds or used in thin layers in compost heaps. Acid-loving plants such as blueberries and potatoes don't appreciate wood ashes. Never leave ashes sitting in the rain as the potash leaches out of them very quickly and wet ashes are almost impossible to spread.

Do not allow ashes to touch plants. However, slugs and snails don't like to cross them so they can be spread in a circle around seedlings for protection.

Some briquettes may contain elements like arsenic or toxic quantities of sulphur and iron. Without testing each batch of briquettes you use you can't tell, so it's safest not to use briquette ash.

Avoid wood ashes from chemically-treated or lead-painted woods. You should also avoid breathing the fumes from these ashes.

Don't use the soot that builds up in your chimney or flue. Apparently it has a very high pH and contains creosote which may or may not be carcinogenic. Why take chances?

Even city gardeners should be able to find friends or neighbours who have a wood-burning stove or fireplace and who will be pleased to have their ashes removed for them. Look out for bonfires and fires for burning off. These ashes are good to collect so long as you can get them before they are rained on.

Worm castings

Worm castings are the bodily waste products of worms: worm manure. If you have ever collected a pile of horse manure from the paddock after it has sat there for several weeks, you will probably have noticed a neat little pile of lovely, crumbly, fine soil. This is a pile of worm castings and it is every bit as good as it looks.

Since it has travelled through the digestive system of worms, worm castings are bound to break down quickly. Don't expect to cover weeds, though. Worm castings are very fine, excellent for use around seeds and tiny seedlings. They make a great soil conditioner, encouraging good aeration and penetration of roots and water, and are usually high in calcium, carbon, magnesium, nitrogen, phosphorous and potassium. They are neutral in pH and can hold up to ten times their weight in water.

Worm farmers produce worm castings as a by-product of growing worms for fishermen and gardeners. Most include leaflets about growing your own when they sell a bundle of wrigglers. You can manufacture castings quite easily by seeding worms into contained piles of manure which have already composed for about a month. Fresh manure gets too hot and kills worms.

Worm castings contain large numbers of worm eggs, so they can also boost the worm population under your mulch.

Alternatively, you can buy a worm 'farm', sit it on your verandah and turn garbage into compost without the neighbours suspecting.

A town in California is using worms to digest sewage sludge, producing 50 cubic yards of worm compost per week. Is this one of those ideas whose time has nearly come?

If you'd like to give the idea a go yourself, there is detailed information about making worm compost in *The Reverse Garbage Garden*, pages 60–2.

FREE AND NEARLY FREE MULCHES THAT DECAY MIDDLING FAST

Agricultural and industrial by-products

This heading covers a multitude of sins. A lot of very diverse materials are included here, so I've taken the easy way out and decided to put them in my middle category.

Often by-products are specific to areas and may only be available from the processors. Sometimes they are marketed commercially. You may have to search for them.

They include apple pomace (available in orchard areas), bagasse (from sugar cane) and brewery waste, made of barley and hop residue, which is used by market gardeners in Tasmania. Whey can be available in dairying areas but should be used sparingly because it has a very high pH.

Euca mulch (see Chapter 6), the residue left over from extracting eucalyptus oil, is a by-product sold through the shops.

There are bound to be many more that I have never heard of.

Bracken

I have included this ancient plant separately because of its abundance on many farms. It is a pest right throughout Australia and can be toxic to stock, as well as highly inflammable in a bushfire. There is some suggestion it may be linked to cancers in cattle and possibly humans via cow's milk. It is difficult to get rid of and spreads from underground rhizomes as well as via spores, but is rich in potash and organic matter.

I mulch with bracken because we cut it frequently on our one-and-a-bit hectares. Regular cutting weakens it. It is a little difficult to apply unless shredded, but unshredded bracken still makes a good coarse mulch around large plants. Shredded, or partially decomposed, it is excellent around smaller seedlings, or for covering seeds.

Shredded bracken decays very quickly into a fine composty mixture. Unshredded material takes longer. It starts off green but quickly becomes a faded brown.

Fish wastes

Fish waste is high in nutrients, especially if it comes from salt-water fish. We all learnt in school that the Indians taught early settlers in America to plant a fish (dead, one supposes) under corn.

Fleshy bits and the skin of fish break down very quickly, but the bones take a lot longer. Small amounts of fish waste can be put under other mulch, such as hay or grass clippings. It can also be used in compost heaps or dug into the ground. Like feathers, fur and wool, it tends to be smelly and to attract flies and cats and dogs.

An alternative with all the animal wastes that smell and bring flies, such as bones, meat scraps and fur, is to circulate them through the chook pen first. Maggots are good feed for chooks, being high in protein. The chooks also pick off any scraps of fish and meat and you end up with clean mulch. You could try processing fish and meat scraps with wood ashes as suggested for bones. I don't know if it would work.

You may be able to arrange for a commercial fisherman or a processing plant or a local fish shop to give you loads of fish wastes.

In city gardens, fish waste should either be avoided or dug deep enough to avoid attracting neighbouring animals. The same nutrients are available if you use commercial fish emulsion fertiliser (at a cost, of course).

Human and animal hair

A household generates only tiny quantities of this nitrogenous material but you shouldn't waste it by throwing it into the rubbish bin with the plastic bags and empty tins. It is fibrous, but breaks down more rapidly than wool or feathers because it is very fine.

You can put it around seedlings and plants but cover it with heavier materials to stop it blowing around. Don't expect your son's hair clippings to kill the weeds, unless your son is a lot hairier than mine! Hair is a mulch that is best combined with other materials.

Hairdressers are a good potential source of hair in larger quantities; if you get large amounts, put it around shrubs and trees. I have had no luck with obtaining hair from my local hairdressers. They objected on the grounds that it was simply swept up with the general rubbish and therefore mixed with too much other stuff. You may get on better with the hairdressers in your area.

Try the local poodle parlour for cat or dog hair. Ponies generate a lot of hair in spring. We get far more horse hair from our two ponies (it comes out in great lumps) than we do human hair from the five family members. There are possibilities with goats, rabbits and the like. If you can get it, use it.

Packing materials

A wide range of materials are used to cushion items sent through the post or by courier. Some of these are nasty little polystyrene lumps that I send to the tip. However, natural materials are also used, often in large quantities. These include shredded paper which is an excellent mulch (see page 53).

Recently I received a parcel packed with a material I had never seen before. Luckily there was a note explaining that, although it looked like polystyrene, it was actually an all-natural plant-derived starch product which would dissolve within half an hour. I put a couple of bits in a cup of water and within a few minutes saw them turn to mush.

Material like this is excellent for mulch, but of course, once there's rain, it disintegrates super quickly!

Take care not to use treated pine shavings or sawdust. You may have to do a bit of detective work sometimes to find out if a packing material is safe to use.

Paper and cardboard

Trees give us bark, chips, sawdust, shavings, wood ashes and also

Grass, hay and paper are conserving soil moisture for this grapefruit plant.

Having well-rotted mulch to sow and transplant into makes gardening easy.

ttuces and onions thrive without atering because of the mulch.

Sawdust covers weeds well, especially with newspaper underneath it.

air is a fine, fibrousy mulch, varied in lour. The dark hair is my husband's!

You don't have to kill birds to mulch with feathers—these were picked up and they will retain their shape and colour for ages.

Faded hay is a light, bright, interesting mulch.

Pine needles are fine and present varying shades of brown.

Shredded bark is light, varying from fawn to browns.

Eucalyptus leaves include delicate and varied pastel shades.

Peanut shells make an attractive, bright mulch.

Seedpods from my local wattles are a rich red—they take months to fade to grey.

Native wood chips are incredibly varied—these are lively light browns . . .

while these native chips are interesting shades of grey and grey-brown.

paper. Paper products undergo a great deal of processing, much of which involves the use of heavy metals or dangerous chemicals and some gardeners are reluctant to use cardboard and paper at all because of possible heavy metal contamination. However, research indicates there's unlikely to be a problem even with very heavy use.

The soil is a terrific filter for dangerous substances and plants rarely take up heavy metals in dangerous quantities. I use paper and cardboard on my paths and mixed with other material around ornamentals, but not in quantity among vegetables. Paper is low in nutrients but disintegrates slowly if put down in sheets or slabs and especially slowly if it stays dry.

Very glossy and highly coloured paper, such as that found in magazines, should be used in moderation. You're only likely to have moderate amounts anyway!

Newspapers are readily available. You can get as much of them as you want. Envelopes, kids' drawings and school work, letters and junk mail are other sources.

Offices shred paper routinely now. If you can establish a regular supply, so much the better for your garden. Cheap, hand-operated shredders are available overseas but not, I think, in Australia. Maybe if gardeners started clamouring for them . . .? Once wet, shredded paper tends to mat down and not blow around. It disintegrates faster than unshredded paper but still gives good weed coverage. Lilian from Mount Colah (see page 124) uses shredded paper.

A Western Australian gardener tells me that he soaks newspapers overnight then puts them through his chipper/mulcher along with sawdust and kitchen wastes. He says it does wonders for his garden.

Paper is useful as a mulch on steep slopes. To stop water running straight down from top to bottom, lay the sheets from the top downwards so that each sheet overlaps the sheet above it. That tip came from another Western Australian.

Cardboard is just as easy to get hold of. Shops throw out cardboard boxes all the time. Bring your shopping home in boxes instead of plastic bags and ask friends and shops to save them for you.

Peat moss

Peat moss (sphagnum) doesn't contain nutrients, but it improves drainage and helps to aerate the soil. It can hold ten times its own weight in water. It is especially useful as a fine mulch covering seeds and around seedlings.

It is made up of the partly decomposed remains of plants and is often found in highly acidic, swampy areas. We have patches of sphagnum growing in badly-drained parts of our place and I have come across springy fields of it in the Kosciusko National Park.

You can try cautious harvesting of your own sphagnum moss if you have it but remember it grows slowly in conditions that are very difficult for plants so don't take much. I have no intention of pulling up any of mine. It is also available commercially.

Pine needles

These can be an excellent mulch, although nothing much seems to grow through them in pine plantations. They are middling good at smothering weeds, best when combined with other materials.

Pine needles are highly acidic so are at their best when used around plants that appreciate the acidity. Blueberries are a good example: they are impossible to grow if the soil is *not* acidic. All berry plants like pine needles. *The Reverse Garbage Garden* gives details about more fruit and vegetable plants that appreciate an acid soil and mulch.

If pine needles are being used around plants that do not demand an acid soil, they can be sprinkled with lime or wood ashes to counter the acidity.

Don't apply them in thick layers or the needles can mat down and prevent rain getting through; several thin layers are better. Pine needles can be applied more thickly if they are mixed with other material such as grass clippings or hay.

Plant wastes

These are a source of innumerable minerals gathered from all over the country. Banana skins bring nutrients from the soils of Queensland, orange peel perhaps from the Murray River area, while apples could have sailed up from Tasmania.

Naturally, being so diverse, these wastes decay at different rates. Banana skins disappear quickly while orange peel sits around looking terrible (unless covered) for a long time.

Do not put vegetable and fruit wastes on top of the ground during summer, as flies will certainly breed in them. Cover them heavily with other mulch, dig them into the ground or the compost heap or feed them to animals.

They decompose more quickly if they are shredded, which can be achieved by roughly hacking them about with a hoe or a spade or by putting them through a mulcher. Manure or urine also helps to break them down. You can put small amounts through a household meat grinder or blender but this could become tedious. Many chipper/shredders will mash up plant wastes.

Plant wastes can be obtained in small amounts from your own kitchen or in larger quantities from fruit and vegetable shops and markets. Ask if you can collect their vegetable waste and explain why you want it. I pick up rubbish from both the greengrocers in my town, although I only shop at one of them. This helps feed our chooks and rabbits and whatever they don't fancy goes onto my garden.

Another source of plant wastes is florist shops. Florists trim leaves and the ends of stalks from the flowers they sell. You can also try flower growers since they have to trim and pack their own flowers.

If you make a commitment to remove wastes regularly, make sure you stick to it or you will risk losing your source.

Naturally, your own waste should go straight back on your garden. When you pick a cabbage, simply throw the outside leaves around other plants. If you pick flowers, trim off unwanted leaves and let them feed the plants that gave them to you.

Tidy leaves and stalks can go anywhere; mushier wastes had best hide their light under a bushel (of grass clippings or straw).

Slag

Slag is one of the mythical materials I've read about but never come across. It's a by-product of the process that turns iron ore into pig-iron, and it's a good source of calcium, phosphorous and plenty of trace elements. It is alkaline, so can be useful on acid and heavy soils.

Although slag is available to American gardeners commercially, in Australia you would have to try approaching industry sources and see if you can come to an arrangement with them. If you live near Newcastle, Wollongong or Point Pirie it's worth a try.

It doesn't sound an attractive mulch, more a useful one, so bark chip manufacturers can sleep easy.

If you use it, spread it thinly and cover with other mulches.

Straw

Straw is the stalk part of plants and provides valuable organic matter. It is more often bought than obtained free.

It is a good, coarse, seed-free mulch suitable for use around larger seedlings and plants. It is attractive enough to be used anywhere and has a bright, clean look. Its colour doesn't fade as rapidly as grass and hay. Straw seems to last for about a season before it needs replacing.

If it is to smother weeds, it has to be laid down very thickly or combined with other mulches, such as paper and cardboard.

Specific straws are regionally available, like pea straw in Tasmania. Straw from legumes, like alfalfa and peas (see page 72), is higher in nutrients.

Weeds

By weeds I mean plants that have been pulled up rather than cut. The tops, leaves, roots and the soil or mulch clinging to the roots all count as weeds.

Healthy green plants always contain the elements necessary for growth: calcium, carbon, hydrogen, iron, magnesium, nitrogen, oxygen, phosphorus, potassium and sulphur. When mulching and composting with weeds, you are returning many of these elements, as well as the organic matter, back to the soil. Chinese farmers have long sought out weeds for their compost heaps.

Weeds can be applied directly as a coarse mulch around large plants and are effective in smothering the weeds they cover. The weeds you throw on the top of the mulch tend to re-grow. When they do, they can be pulled up and inverted or simply covered with more mulch.

Don't use weeds which grow from underground parts of the plant and spread readily (see also page 39). I have finally been forced to recognise that sorrel is not suitable as a mulch, since it spreads under mulch to become a pest. The only time I tried to mulch with couch grass, I was forced to pull it out of the garden and throw it onto the driveway. Avoid nutgrass and oxalis, too, although weeds like these can be used if they are first thoroughly dried out and killed, or composted at a high temperature (see *The Reverse Garbage Garden,* Chapter 8). You can use the leaves, but be careful to exclude the stems.

You should also be careful if you decide to mulch with noxious weeds—be responsible. I received advice from a Department of Conservation and Natural Resources officer (but it may have been off-duty advice) that I could use ragwort so long as I made sure it was completely covered by other material. Ragwort is a declared noxious weed in Victoria. Not only does it seed profusely, but the seed continues to mature after you have pulled the plant. This makes it dangerous as a mulch.

Weeds are not the most attractive mulch—use them to nourish the soil rather than for landscaping.

Hugh from Mount Colah has an interesting and opposing point of view on the use of green weeds as mulch. (See page 124.)

FREE AND NEARLY FREE MULCHES THAT DECAY SLOWLY

Bark

Bark is rich in minerals. A shredder/composter can shred it and it can then be used anywhere in the garden where it will decompose fairly quickly. Unshredded bark is difficult to handle and bulky. It is not suitable for putting around seedlings and seeds, but it makes a good mulch for larger perennials, shrubs and fruit trees and decays slowly.

Bark has to be applied in thick layers or combined with cardboard or paper, or there will be spaces between the slabs or strips for the weeds to push through. Forty centimetres is not too deep for this mulch.

Slugs and snails love slabs of bark, which make good snail traps. Just pick up the pieces of bark and destroy the pests that are clinging to them!

Bark looks better than you'd think and decays within a year or so.

Cloth and carpet

Natural fibres make good mulches for the perennial sections of your garden. They tend to get in the way with annuals. They are not high in nutrients but contribute useful organic matter and block weeds if put on thickly. Old carpet is especially good for suppressing weeds that are extensive and hard to handle, like blackberries.

Old towels, clothing and smaller items like face washers and tea-towels can cover large areas quickly but are not attractive-looking mulches. Cover them with neater mulches in places where appearances count (your front yard, perhaps!).

Do not use synthetics or clothing sewn together with synthetic thread, as these will not decay and can ruin brushcutters and lawnmowers. They can even make it impossible to handpull weeds.

Old underfelt or underlay from carpets can be obtained from friends or from contractors and factories that lay new carpet. It can be used on large areas or can be cut into small sections and positioned around individual seedlings. Cut a hole in the middle for the plant. Blood and bone or manure can be laid down first. The underfelt can easily be removed at the end of the season if it hasn't already rotted.

One word of caution about underfelt—when it is used under carpet it is sometimes impregnated with dieldrin to protect it against insects.

Felt from old clothing, craftwork and hats is another versatile material, good around flowers, fruiting plants and even vegetables.

If, like Jeff from Argenton (see page 123), you can get hold of a supply of old hessian bags, they make a brilliant mulch. Lay them down thickly or put paper underneath to stop weeds growing through the loose weave.

Feathers, fur and wool

They sound like a children's television show! All these materials are high in nitrogen. They are also slow to decompose and awkward to use in the garden.

Small quantities, such as you would get from killing your own meat, can be put under other mulch or around trees or dug into the ground. If there is offal to dispose of at the same time, dig the whole lot under at once. These are not materials you're likely to use in your front yard in inner Sydney unless they are well covered.

Farmers, processers and slaughterhouses may give you access to larger quantities. Because they are slow to decay, large quantities are best put under or around shrubs, trees and vines rather than in with the annuals.

Bits of wool impregnated with urine and manure are often available from shearing sheds. These rot faster because of the nitrogen. Lilian from Mount Colah (page 124) has a friend who mulches extensively with lumps of wool from discarded fleeces and Robyn (page 128) says that wool dags are often available from the tip near her on King Island.

These materials may attract cats and dogs who can make a mess of gardens by scratching and digging. Covering them with other mulches helps but, if you have a small garden, it's best to avoid them.

Leather scraps

These can be anything from a mouldy sheepskin you intended to cure to offcuts from your kids' artwork.

Leather is processed animal skin so it contains nitrogen and organic matter and undoubtedly other things to contribute to the garden. Either use large pieces whole for their weed-covering capacity (among perennials please) or snip them into small pieces to speed up decay.

Your mouldy sheepskin that you always intended to tan properly when you got the time will make good mulch, but clean the blood and fat off it if you don't want to breed maggots. Slugs and snails will like the large pieces, so treat them as snail traps. Pick them up and check underneath regularly.

You can get leather offcuts from artists, craftworkers and factories or old bags, belts and clothing from Op shops. If you come across any great leather jackets, just let me know *before* dumping them on the garden!

Leaves

Leaves provide large quantities of fibrous organic material, which breaks down slowly adding humus to the soil.

Leaves tend to blow away, especially the large leaves of deciduous trees. Spread a thin layer of sawdust over them to hold them in place. If the leaves are shredded before use, they break down much more quickly, or they can be mixed fifty/fifty with grass clippings. Blood and bone, diluted blood, wood ashes and urine all help leaves to decay.

Leaves are a source of minerals such as calcium, magnesium, phosphorous and potassium. Sprinkle lime or wood ashes on leaves to correct their acidity unless you're using them around acid-loving plants.

Leaves are abundant in the city as well as in the country. Autumn sees huge mounds of deciduous leaves banked up on footpaths, against fences and trees and in parks. Gather them in bags, shopping baskets, old prams and car boots. If you don't need them straightaway, keep them in plastic bags and use them throughout the year or turn them into leaf mould (see pages 45–6).

Native leaves can be gathered from under trees at any time of the year. Wattles and many other trees and shrubs give excellent leaves and leaf mould. Don't apply large quantities of eucalypt leaves, since they may contain chemicals that inhibit the growth of plants. Shredding helps, as does partially composting them.

Plastic

I have never used plastic on the garden, but that's not a good reason to ignore this material!—I've never used iron or straw either.

Many gardeners swear by commercial plastics, so why not use waste plastic—heaven knows, there's enough of it around. In fact, take out the sixty-five per cent of your rubbish that consists of kitchen and garden wastes and paper (all of which are now going onto your garden) and the proportion of plastic starts to look considerable.

Plastic won't supply nutrients or organic matter to the soil, but it will cover weeds, stop moisture loss and warm the soil. If it looks ugly it can be covered with a more attractive mulch. Modern garden plastics let rain through, but waste plastics won't, so be careful that plants don't dry out under their mulch.

Clear plastic has an advantage over the once ubiquitous black. It warms the soil more. However, weeds can grow under it because of the light coming through and the warmth—cover it with mulch or weigh the plastic down to stop the weeds pushing it up or finding their way through gaps between pieces of plastic.

Plastic can be laid around plants, or you can cut holes in large sheets for the plants to grow through. Remember that it can't supply *any* nutrients—it's best to see it as a temporary way of suppressing weeds. Replace it with a richer mulch when you can.

Sawdust

Sawdust is very fine, easy to apply and has good covering capacity. A layer just 10 centimetres deep stops weeds getting through. It is

very low in nitrogen and doesn't rate as a fertiliser. As sawdust rots and gets pulled down into the soil, it aerates the soil and increases its water-holding capacity.

Sawdust is my preferred mulch for paths, because it effectively suppresses weed growth and air-borne seeds are unlikely to germinate in it. It is, itself, seed-free.

It is also effective as bedding for animals. It soaks up urine and stops pens, sheds and yards smelling. It can then be put onto the garden as a mulch with high fertiliser value.

Sawdust is free if you pick it up from sawmills, or it can be delivered at a cost. Carpenters and various manufacturers, such as those making furniture and packing cases, also create large quantities of sawdust, as do the woodworking departments of schools. Mostly it's available free if you take it away, as they are pleased to get rid of it. It can be collected in different quantities, from a single shopping bagful right up to a truck load. Avoid sawdust from treated pine products. If you're unsure, just ask.

Sometimes well-rotted sawdust is available because unwanted material has been left to weather in a heap. It is dark brown and crumbly—a fine soil conditioner and a terrific mulch to put over a row of seeds or around newly-emerged seedlings.

Sawdust is an attractive mulch that can be used anywhere. Initially light, it mellows through a deep reddish-brown to black.

Seashells

Instead of selling seashells on the seashore, we can harvest them for the garden.

Seashells are mostly lime so don't use them if your soil is alkaline. They are attractive, but difficult to obtain in any quantity. Broken shells are preferable to whole ones, or they can be crushed and sprinkled over the garden. Uncrushed, they are awkward to handle in annual beds so put them around perennials. An under-mulch of paper or plastic helps to keep the weeds down.

Before collecting shells or seaweed, check that you are allowed to do so.

Seaweed

Seaweed is high in trace elements and potash. It is suitable for use around larger plants and some gardeners claim that the minerals in seaweed make plants resistant to frost damage and prolong the harvest of autumn tomatoes.

There are green, brown and red seaweeds, all of them good to use. Soft, fleshy seaweeds rot fairly quickly (kelps, for example) while seagrass rots very slowly. I still find bits of seagrass eight years after putting it on the garden! Smaller seaweeds are easier to handle than the enormous bull kelp.

There is no need to wash the salt off seaweed, especially if it is just one among many mulches. You can let it dry out, then run over it with a lawn-mower to shred it, thus making it a finer mulch and one which breaks down more quickly. I don't know if you can get it through chipper/mulchers.

Seaweed is absolutely seed-free, doesn't harbour any plant diseases or insects and looks much better than you think.

Ruth from Bell Park (see pages 126–7) mulches exclusively with seagrass and can't praise it highly enough.

Shells and hulls

All hulls and shells are suitable for mulch, either on garden beds or around flowers, shrubs and trees. They must be applied thickly to suppress weeds—at least 10 centimetres deep or combined with other mulches.

Most shells are slow to decompose but eventually contribute organic matter to the soil. They make clean, non-smelling mulches that don't normally harbour maggots. The one problem I've found is that slugs like taking shelter in the broken nut shells. It's worth checking the mulch now and then for this.

The small amounts you get from an ordinary household can simply be put on mixed with other mulches. Large quantities may be available from farmers, food processers and confectioners. Almond, hazelnut, peanut and pecan shells are worth looking out for. Walnut shells contain a chemical that inhibits growth but, in small quantities in mixed mulches, there will not be a problem.

Rice hulls are commercially available. A friend gets his free from a farmer's dairy where they have been nicely mixed with manure. Rice hulls are very slow to decay, because of their very high silica content—they are a good soil conditioner and a tidy, long-lasting mulch.

I haven't heard of other hulls being used, but commonsense says there must be husks from oats, sunflower seeds, wheat and many others kicking around somewhere. Perhaps somewhere near *you*.

Jeff from Argenton (see page 123) considers cottonseed hulls or husks to be a top quality mulch.

Twigs and branches

These are similar to the stems of plant wastes but are bigger and harder. They may consist of prunings from your bushes and trees or waste stuff you've chopped down, or come from plants especially grown for providing mulch. I wouldn't put them in the front yard because there are much tidier mulches.

Gardeners with chipper/mulchers are already accustomed to using this material. They shred it into coarse or fine chips to make a terrific looking mulch. If, like me, you don't have a machine, you can break it up a bit by hand and lay it around your bigger plants. Don't try to put it near seedlings but it's fine for well-established plants, shrubs and trees. The larger branches can go around the larger shrubs and trees.

Leaves are very rich in minerals so this mulch has nutrients as well as organic matter to contribute to the garden. If you collect your own firewood, you are likely to end up with the small ends of branches that are not worth burning—here's a use for them.

Branches from legumes, like tree lucerne, supply even more nutrients (see pages 85–7).

Don't expect the weed coverage to be very good unless you combine twigs and branches with other material like hay, paper, etc.

FREE AND NEARLY FREE MULCHES THAT TAKE A MILLION YEARS TO DECAY (ALMOST)

Concrete slabs

This is not my own idea and I have never seen it done. It is a variation on stone mulching that I read about in an early copy of the American magazine, *Organic Gardening and Farming*.

They use a pool ball rack (pool being that queer game some people play inside with chalky sticks on green tables) as moulds for making triangular-shaped concrete slabs. These fit together much better than natural rocks to make a neat, long-lasting, warmth and moisture retaining mulch.

Of course, the blocks can be of any shape or size, but keep them small enough to handle and flat on the bottom.

Crushed rock, such as dolomite, can be added to the concrete mix as a slow release fertiliser.

The slabs can go around perennials or on paths among annuals. If you put cardboard, paper or plastic underneath them, you will make sure that weeds don't get through. You can also combine them with mulches like grass clippings, manure and seaweed if you want to supply plants with plenty of nutrients.

Don't use them around acid-loving plants since cement is basically limestone and therefore alkaline.

Metal

This may seem a strange material for your garden and certainly it wouldn't do in city areas. However, if appearance doesn't matter and weed suppression does, old corrugated iron can be useful. You'll already know this if you've ever had to clean up a messy farmyard. Depending on how long the metal has been lying around, there'll be nothing underneath it except pallid, dying grass or barren desert.

Sheets of metal are good on problem weeds like blackberries. You can cover large areas, treating them as stage 1 in your land reclamation programme. Stage 2 is to remove the iron and plant out the area once the weeds are under control.

The metal smothers weeds (put paper or plastic under holes in the iron), protects the soil and leaks small amounts of minerals. It also stops rain getting through, adds no organic matter and has to be weighted down. Use stones or slabs of wood.

This is a rough-looking mulch and difficult to handle (sheets of iron are surprisingly sharp along the edges and have caused many a nasty cut), but it is useful in the right circumstances.

Don't try to shift sheets of corrugated iron around in windy weather and keep them weighed down so they won't go flying in some minor gale. Make sure kids (and gardeners with bare feet and ungloved hands) are not likely to damage themselves.

Stones

Most gardeners think of stones as obstructions that have to be grubbed out of any self-respecting garden. They can, however, be used as mulch, especially around perennials or in the paths and along the edges of annual beds.

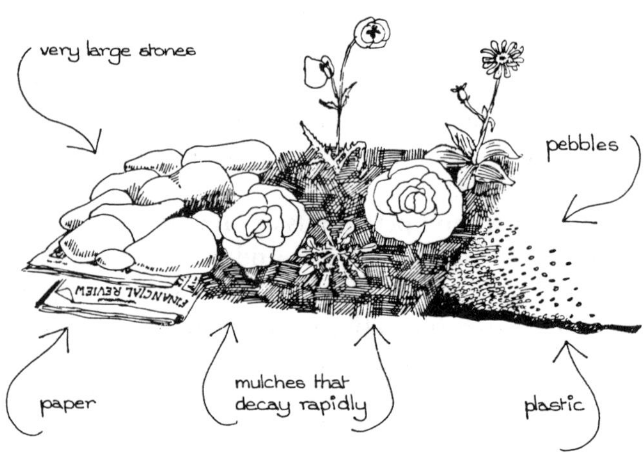

Used as mulch, stones reduce moisture loss from the soil, reduce weed growth, store solar energy and slowly release minerals into the root zone of plants as rain leaches them out. They are free and readily available in many areas. On the other hand, they are heavy and awkward to shift around and they don't decay rapidly (to say the least).

Because stones absorb and retain the sun's heat, they are particularly useful during the cooler months. In summer, it may be necessary to cover them with other mulches, like grass or leaves, to reduce their heating capacity.

Plant roots may actually grow into stones and rocks, absorbing some of their mineral content.

For good weed control, stones should be combined with other mulches, such as paper and cardboard.

Stones can refer to anything from pebbles to lumps just light enough to carry. Large stones should be laid close together and the holes between them filled with gravel, leaves, sawdust or similar material. The flatter the stones are, the easier they'll be to lay and fit together. Whether large or small, stones make an attractive and permanent mulch. Of course, they don't add organic matter to the soil but they do help look after it.

See also Concrete slabs.

Wood

Planks and bits of wood can be used as mulch. Small and rotten pieces disappear very quickly if they are kept moist. As the wooden sides of my garden beds decay, I throw them in with the rest of the mulch, but I don't use wood around tiny seedlings.

However you *can* put planks on top of seeds. This conserves moisture and encourages better germination. You must check them daily and remove the wood *as soon as the seeds start to germinate.* Watch out for slugs, too. Planks or slabs of wood make good traps for slugs—just lift them up regularly and kill what you find.

Large, discarded bits of wood can be used to cover weeds. Make sure they overlap heavily or combine them with other mulches, or the weeds will come up through the gaps. As they're not exactly attractive, you'll probably want to cover them with grass, hay or the like anyway.

Avoid treated pine wood because it has been soaked in chemicals which are designed to kill insects.

Although it can't supply much in the way of nutrients, if it is kept moist wood rots reasonably quickly and supplies organic matter.

6

Commercial Mulches

Believe it or not, it is easier to find out about free mulches than commercial mulches. God knows why. Many companies making good money out of mulch promised me information; almost none followed up.

Don't they *want* to sell the stuff?

My information is sadly patchy but I'll do my best.

Commercial mulches fit into several more or less neat categories:

1. wood products;
2. stones;
3. locally-produced mulches;
4. organic mulches (that is, they are marketed as organic; there's nothing inorganic about stones, after all!)
5. industry by-products;
6. weed mats.

WOOD PRODUCTS

My local contractors gave me a guided tour of their operation. They sell bark and wood chips plus euca mulch (see page 71).

These guys have a yard that looks like it dropped in from outer space. There are strange machines and big tubes and piles of . . . something . . . and motors banging away. Their machines are crushers and grinders and cutters allied with tubes made up of rotating screens of various mesh sizes. Bits fall through the screens when they are small enough and make up the towering piles of different sized lumps.

There are also immense piles of raw sawdust.

They sell mainly pine wastes—bark, shavings and chips of wood left over from milling timber at the (fairly) nearby APM (Australian Paper Mills) works. They get the waste in its raw form and put it through the torture chambers until the bits measure under 10, under 25 or under 50 millimetres. Local gardeners use the two finer grades: 10 millimetre material starts to decay after six to twelve months. Under 50 millimetre chips are used by councils, etc., where cheap material is wanted in bulk.

Some of the under 10 millimetre pieces go through a vibrator which shakes them around and separates them. This results in a super fine under 4 millimetre mulch, which is used in the nursery industry as a potting mix, as well as a grade between 4 and 10 millimetres.

Shavings and flakes from squaring off and planing pine logs are mainly used as a cheap mulch in playgrounds and parks.

Basically, the finer the grade of chip, the more expensive it is and the faster it tends to decay. Sawdust and flakes, however, are very cheap.

The local contractors occasionally have red gum chips, but prefer pine because they get it raw and process it themselves. The red gum comes ready made from north-west Victoria where the timber is milled.

All of their material is a waste product of the timber industry and trees are not cut down especially in order to provide it. Their ability to make full use of our timber resources is to be encouraged and applauded.

Bark and wood chips in different sizes, flakes, shavings and sawdust are widely available throughout Australia. The particular type (pine, red gum, etc.) depends on the wood that's milled in your area and how the contractors operate. Some just truck stuff in from interstate and sell whatever's going. Others, like my local contractors, prefer to sell their own value-added product.

Never use wood-based products that have been made from treated wood. Just ask the contractor, sawmill or whoever you intend to buy it from. Material marketed specifically as mulch is unlikely to be made from treated wood but, if in doubt, check it out.

Wood products are widely available at reasonable prices and can be bought in small or large amounts and delivered. They are long-lasting and low on nutrients; not suitable for extensive use in vegetable gardens except for mulching pathways. They are good for Australian plants and permanent plantings and make good weed covers. They are normally used with paper and cardboard or plastic laid underneath. The chips, flakes and sawdust are effective in inhibiting seed germination for a year or two.

I know many gardeners who use fast decaying free mulches that are high in nutrients for their fruit and vegetable gardens out the back, but wood chips on top of plastic for their shrubberies and flowerbeds in the front yard.

Termites may be attracted by wood products (see pages 113–14) so be careful about using them if termites are a problem in your area.

Wood based materials are relatively light to transport and shift around, and are attractive. If you get sick of them being in one part of your garden, you can always up and move them (it's a touch harder with stones!). A range of colours and textures is available from the rather flat yellow of fresh sawdust to the rich and varied reds, oranges and browns of bark and wood chips. Shop around and get the look that suits your garden.

STONES

Pebbles and stones are available in a range of sizes from miniscule to fist size. They can be obtained from nurseries and landscape suppliers.

Stones come in a vast array of colours and textures. White and pastel-coloured pebbles were used over plastic around the library in a nearby town. They measure about 2 centimetres in diameter and lie maybe 10 centimetres deep. They are highly reflective (downright glary in sunny weather), but attractive where there is a lot of greenery to soften the glare.

A friend of mine has scoria in her front yard. This is a porous, volcanic rock, pockmarked with air holes and very lightweight. The pebbles, 3 to 4 centimetres in diameter, are red. I wouldn't call it exactly attractive but it does its job and is easier to shift around than heavier pebbles. The colour is rather crude and monotonous.

The same friend has blue metal gravel in her backyard. They are blue-grey basalt chips, between 1 and 3 centimetres in diameter. The colour is pleasant and soothing but I feel as if I'm walking on a road or a driveway rather than through a garden area.

These three examples are a fraction of what's available. Look around. Visualise the material in place—are you going to get an effect you like or can live with?

Stones keep the soil cool, retain moisture and slowly release minerals that can help plant growth. They are not particularly good at suppressing weeds unless plastic or paper has been laid underneath them.

Except for scoria, they are heavy to shift around. Very heavy!

Left: Wattle seedpods lie thickly beside roads and driveways in season.

Below: These fine pebbles have naturally varied colours.

Wood chips can often be bought or obtained free from contractors.

This sawdust will go onto the garden after serving as bedding for our turkey couple.

From left to right this potato plant is surrounded by long grass, horse manure, hay, grass clippings, sawdust and leaves.

Grass and hay keep weeds away from young onion plants.

Fresh grass clippings are an excellent mulch and high in nitrogen.

Grass clippings are a gc

Straw is a lovely, bright garden.

Bracken is a coarse mul content is sure to bene

:h for small seedlings.

This young tomato seedling has been planted into pea straw mulch.

ch in this mixed

Loquat leaves are mulching young broad bean plants.

t its high potash
young passionfruit.

The densely planted raspberries stop the newspaper mulch from blowing around.

Leaves are used to mulch this indoor plant.

These daffodils are thriving in wood chips made from Australian shrubs and trees.

These plants are rapidly getting to the point where they won't need mulching.

Once they're in place, you're unlikely to want to move them. Their weight makes transporting them relatively expensive.

Apart from landscaping, stones can also be used around berries and fruit trees. The minerals they slowly release will be welcomed by fruiting plants and the stones can be combined with mulches that decay quickly such as grass clippings, hay, manure, seaweed and straw. Very large, flat stones are good around fruit trees.

LOCALLY PRODUCED MULCHES

In many places local councils produce compost or mulch, which is either sold cheaply or is available to residents free.

This is an idiosyncratic development. It seems to depend on the commitment and work of specific employees with an interest, and possibly experience, in this area. There is no central body which plots the hot spots where the material is available. You can't buy a map or a directory. You have to ask around and be observant. Enquire at your local council or ask local gardeners.

Often the basis for these mulches is chipped branches and leaves. The chips may be mixed with other material and left to decay a bit. In some areas organic rubbish, instead of being sent to the tip, is composted or partly composted and often mixed with chipped prunings.

In Melbourne's crowded St Kilda area, the council is composting seaweed. The seaweed has to be removed from the inner suburban beaches because it gets smelly and rubbish gets caught in it. Then it is spread out to decompose. The finished product is being used as a potting mix, for planting on reclaimed foreshore land and in local vegetable gardens.

When they are free, these composts and mulches are obviously good value. Even if you have to pay for them, they are often rich in nutrients and relatively inexpensive. By buying them you are also supporting recycling programmes.

You can count yourself lucky if you have access to locally made mulches and composts. If you don't live in one of the crazy patchwork squares where this stuff is available, why not pester your local council to start a composting scheme? Even if there is no seaweed around (and I have to admit there isn't much in Alice Springs or Wagga Wagga) there is probably plenty of waste like cut grass, prunings and leaves for chipping and household waste. Stir them up (the rubbish *and* the council!).

ORGANIC MULCHES

Wood chips and stones are organic, as are the mulches in the previous section, but by organic mulches I mean those that are specifically marketed as such to organic gardeners.

There are undoubtedly organic mulches that I have never heard of because they are not available everywhere or are not widely advertised.

I did come across an interesting brick wall in this field. I contacted a company which sells organic compost and mulch. They clean grease traps in Melbourne restaurants and sell the material after an arcane and expensive treatment. I say arcane because the guy on the phone became extremely evasive about the whole procedure.

Apparently the conversion of waste into compost is at the cutting edge of business development. His plant cost about $3,000,000 to build and, now that he is about to start pulling in profits, he is afraid of industrial espionage, which he assured me is rampant. It appears it's a cut-throat world out there in organic composts. He told me that the guy who developed and sells Dynamic Lifter has guards patrolling his property!

There seem to be companies processing the waste out of grease traps throughout Australia, however the industry may be most advanced in Victoria. I managed to meet Morrie Caddie of Universal Recycling. First the grease is separated out of the waste and sold for things like soapmaking. After extensive processing, the organic waste is sold as Garden Energiser either in 25 litre bags or delivered in bulk 3 metre loads. It is available by mail order.

Almost everybody has heard of Dynamic Lifter. It's used extensively by professional and semi-professional organic growers, although more as a compost than a mulch. It is a clean, tidy, inoffensive material that you can spread around plants for a nutrient boost, then cover with heavier mulches. The basic ingredient is chook manure so it is designed to give your plants a 'lift'. It absorbs and holds a great deal of moisture.

You can buy Dynamic Lifter virtually anywhere—if I can get it in my local supermarket, *you* should be able to find it. It is available in small quantities or larger.

Seaweed Sales Australia make compost, mulch and liquid fertiliser from seaweed washed up near Kingston in South Australia. Seagold is a friable compost which can be spread thinly as a fertiliser. They also market a fine, ribbony, dried seaweed as mulch. Both are excellent materials, with high nutrient content and good water-holding capacity, and they are available in small or larger quantities.

Worm castings are available all over Australia. They are rich in nutrients and a great soil conditioner. They are also neutral in pH and won't burn your plants. As a bonus they contain worm eggs which, under good conditions with plenty of mulch, will give you a free crop of soil-improving worms! Worm castings are not only excellent as a fine mulch for seeds and seedlings, but they can also be used to make a garden tea for giving your plants a boost (see *The Reverse Garbage Garden*, pages 68–70).

Fauser Worms, situated in South Australia and selling via mail order as well as direct, offer something they call Worm Mulch. This stuff is formulated to provide top quality nutrition for worms. It is made of pea and wheaten straw mixed with chook manure and all neatly pelletised. It can be spread heavily as a weed suppressant and provides nutrients as well as absorbing and retaining moisture.

I recently came across a new product called Multicube Grow 'n' Mulch Cubes. They are made from lucerne hay, cereal straw and 'a complete organic fertiliser'. They are neat, greenish cubes which swell to two or three times their original size when they are wet. They cost $12.50 for a 20 kg bag.

INDUSTRY BY-PRODUCTS

Euca mulch is the by-product most commonly used. This seems to be what is left over after extracting eucalyptus oil. My local supplier puts euca mulch through a hammermill which reduces it to fairly fine particles, but it is also used in larger pieces.

Euca mulch has good covering ability and is attractive. It breaks down slowly, but faster than the bark and wood products. It tends to mat together, which can be beneficial if it is used on steep slopes. It is considered better than wood chips by some Melbourne councils precisely because it mats and stops soil erosion.

It is a good mulch for Australian plants and ornamental plantings; otherwise mix it with higher nitrogen mulches.

Mushroom waste is a rich, composty material, widely available in small to medium quantities. It can be used around seeds and seedlings or sprinkled thinly on the beds to provide a nutrient boost. You can make garden tea from it (see *The Reverse Garbage Garden*, pages 68–70). Dilute the mixture and spray it on plants, or water them with it.

Pea straw, left over from harvesting peas, is used right through Australia from Tasmania to Queensland. I don't know whether it is available in the west. It can be spread thickly for weed suppression and supplies reasonable amounts of nutrients as well as organic matter. It decays quickly. It can be used lightly to cover seeds which then sprout and grow through it.

There are plenty of other materials but a lot of them are not widely available. Sometimes these sorts of mulches can be obtained free (see Chapter 5) if you are in the right place at the right time. Some are marketed professionally through hardware shops, nurseries, supermarkets and other retail outlets. Some are available through local contractors. Use a bit of creativity—look at what local gardeners are using, ask people what's around.

There's been a problem at my local tip with shells from a nearby fish processing plant. There are just too many shells for the tip to handle hygienically. All we need is for some bright spark to figure out how to crush them, and hey presto! we'd have a brand new garden product.

Keep looking and thinking.

Weed mats

PLASTIC

Most weed mats are made from plastic. Some gardeners like plastic for mulch and others won't touch it. I'm in the latter group so I have had no direct experience of this material.

Plastic stops evaporation and black plastic stops weed growth. It absorbs solar radiation. Although black plastic radiates most of this into the air at night, more is retained by clear plastic which, however, doesn't stop weed growth.

One problem with most plastics is that they prevent water and air getting into the soil. The holes that have to be made to allow plants to grow and be watered seem to cancel the effect of the plastic as the weeds grow through them. However, I know one gardener who prefers plastic to newspaper. He says dry newspaper is completely impermeable and the soil ends up compacted and sour. Perhaps I don't have this problem because I'm in a high rainfall area. When I use paper it stays wet and consequently decays quite rapidly.

Plastic is normally used in two situations:

1. to provide a complete weed barrier;
2. to encourage plant growth by retaining soil warmth and/or by weed control.

Black plastic is very effective as a weed barrier. It is usually covered with a long-lasting mulch such as wood chips, which holds the plastic in place and looks good. The combination works well in permanent beds to provide weed control until closely planted shrubs and flowers cover the area and make their own mulch (see Chapter 7).

Plastic also provides a low maintenance garden. Naturally, it has no nutrient content.

Sometimes plastic is used for growing warmth-loving annuals like canteloupe and eggplant, because it warms the soil up and keeps it warm. Nutrients have to be added under the plastic or watered in.

Similarly, plastic is often used for strawberries. Many commercial growers swear by it (whereas I swear *at* it). Weeds grow through the slits made for the strawberry plants and also through any tear in the plastic. The soil underneath has always seemed dry and packed to me. On a backyard basis, I prefer organic mulches like hay, pea straw, pine needles, seaweed and straw.

A minor problem (it didn't seem minor when I was picking strawberries for a wage) is that plastic harbours snakes and red-back spiders in a way unrivalled by the organic mulches which decay instead of staying in place for years.

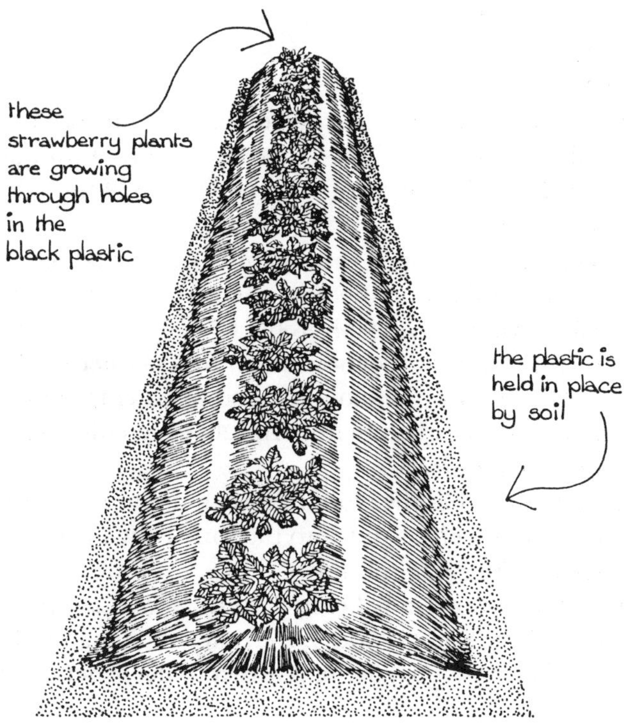

Newer forms of horticultural plastic are superior as they have a woven structure, which lets air and water through but not weeds. They are available in various roll widths or as squares which can be laid around individual plants. The material can be held down by ground staples and is resistant to sunlight.

If you want to use plastic in a big way, go for the new weed control mats that are permeable, so that you can keep the soil in good condition. They are widely used by councils and shires who need low-maintenance, low-fuss gardens.

In a vegetable garden, plastic can be used in pathways or between rows, leaving the earth bare for organic fertilisers along the rows.

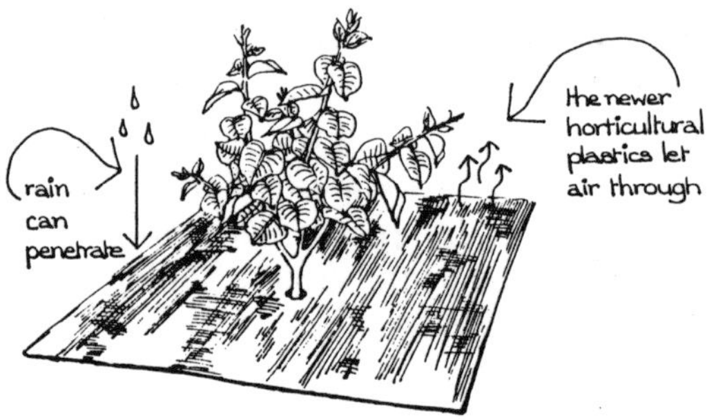

WOOL MATS

There is an alternative to plastic weed mats, an Australian felt product called Ezi-Wool Mulch—there are no prizes for guessing what it is made of! This is a commercial version of underfelt for mulch (see page 58).

Wool mats were developed in England, using good old Australian wool. Now the CSIRO and the Australian Wool Corporation have co-operated to bring out a local product.

The sole distributor at the moment is Tom Garner in Melbourne. He told me that plastics were old hat and wood chips discredited. According to him, plastics are impermeable and chips release acids which inhibit plant growth.

> 'The fertilisers that the wool releases gradually, neutralise the acids that wood chips put into the soil. That's what experiments in England show.'

Ezi-Wool Mats are already being used in large quantities by many Victorian and New South Wales government bodies, but the product is only now (1994) being released to the general gardening public.

It comes in rolls in different weights, from 150–500 grams, with 300 grams being the 'regular' size. This is available to backyard gardeners for $5 per square metre. It's cheaper on a grand scale (almost half the cost of plastics covered with wood chips) and can be installed for you.

You can put it down and make cuts in it for plants, or you can cut out pieces to fit around plants.

It is very effective in controlling weeds and will last at least three years before decaying. Tom says that weeds rarely grow over it, but you can sow your own plants, such as grass or flowers, on top and the roots will range right through the mat.

Unlike plastic products, a wool mat provides organic material and nutrients. It absorbs and retains water, maintains soil warmth and, being an earthy colour, blends in nicely.

If you're not in Victoria, don't panic—Ezi-Wool Mulch is available by mail order.

7

Living Mulches

All organic mulches were once living or came from living things. Cut grass and wood chips clearly were alive not so long ago. Paper and rotting planks of wood may not immediately bring forest (or plantation) trees to mind, but they can be traced back nonetheless. Manure you might prefer *not* to trace back, at least not in person.

Some mulches can hardly be said to have lived—corrugated iron, plastic and stones for example; at least, not unless your definition of living is so broad as to be meaningless.

When I refer to living mulches, however, I mean that they are still living when you harvest them, and maybe the source will live on to be harvested time and time again. Some of them may grow in the wild, either as native or feral plants, others you can grow yourself.

You can even have your living mulch grow right where you are going to use it.

THE TAMING OF THE MULCH

The simplest kind of wild, cut-and-come-again plant is grass. If you gather slashed grass from neglected yards, paddocks, roadsides and wasteland, it's not in the form of neat lawn clippings, but feral grasses. They are likely to be long and weedy and to contain seeds.

This is an excellent mulch for larger plants. It doesn't pack down and contains a wide range of nutrients. The grass, where it was cut, grows again and can be cut continually. Some gardeners value it highly enough to cut it themselves, others wait for the council or shire to do the job.

Weeds can be collected from the same source. The leaves of deep-rooted weeds, like dandelion and dock, are especially valuable since minerals are brought up from deep in the ground and concentrated in the plant. Avoid seeding specimens or make sure the weed mulch is covered by other material.

Be aware of the problem of herbicide contamination. Councils may spray weeds to get rid of them. Ask, if you're in doubt. Weeds that have been sprayed are unlikely to benefit your garden and should never be used near fruit and vegetables.

Avoid noxious weeds or use them with care. Irresponsibility has helped spread pests in the past—don't add to Australia's feral plant problem.

Weeds that are common and useful include: amaranth, bracken (see page 51), catsear, chickweed, cleavers, dandelion, dock, fat hen (*Chenopodium album*), fennel (which has feathery leaves and smells like liquorice), horseradish (if you crush the large leaves between your fingers they have a distinctive horseradish smell), nasturtium, nettle, plantain, sow thistle and watercress.

Try to avoid blackberries, prickly pear and thistles because they are dreadful to handle, even after they have dried in the mulch for months. Nettles, however, rapidly lose their sting and are high in nutrients as well as being edible (see *The Reverse Garbage Garden*).

Many useful weeds will be in the cut long grass but they can also be gathered individually for their high mineral content. Only the above-ground parts are used, so no roots are brought in to regrow. As well as looking along roadsides, you can find weeds on vacant land and railway lines and stations. Some weeds like rich soil and are most likely to be found around farms or where people once farmed or gardened. These include chickweed, fat hen and nettles.

Deep-rooted weeds are especially good sources of minerals, but all weeds are useful. Dandelion, fat hen, nettles and sow thistles are all rich in calcium. Copper is copious in chickweed, dandelion, nettle and sow thistle. For more information about the mineral content of weeds, see *The Reverse Garbage Garden*, Chapter 8.

The simplest form of harvesting is pulling by hand. I do this on my garden edges as it is easier to break off grass and weeds by hand than to fire up our brushcutter. An intimate friend assures me he used to cut his lawn with scissors as a kid. Getting more technologically daring, grass and weeds can be cut with fernhooks, scythes and sickles or, if you're *very* brave, brushcutters and lawnmowers. See also the information about equipment in Chapter 8.

Wild shrubs and trees

You can collect leaves and debris from under shrubs and trees, and branches can also be broken or cut off. Australian plants like acacias, callistemon (bottlebrush) and casuarinas are good, as well

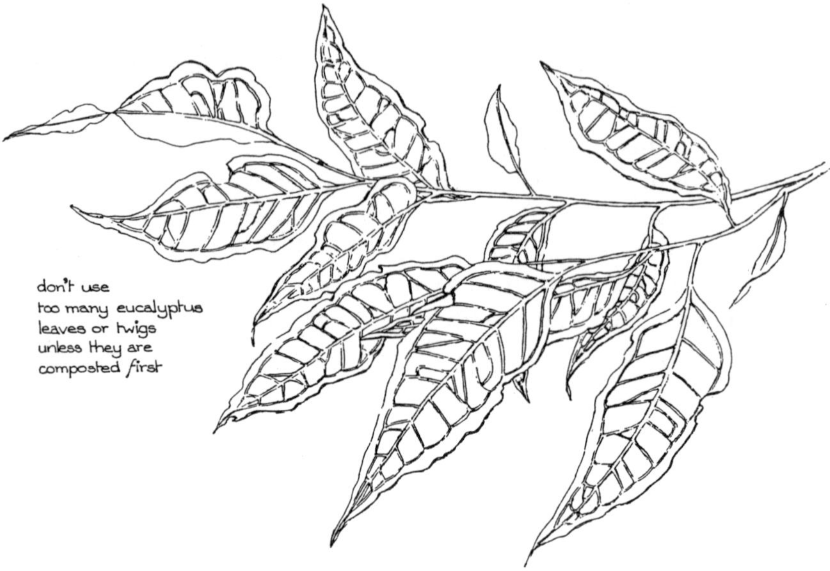

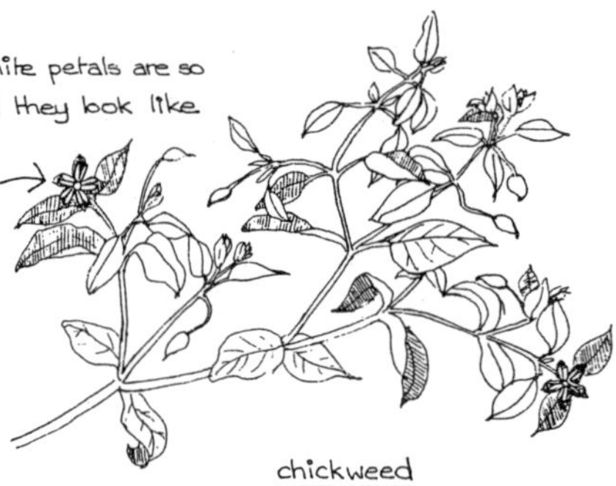

chickweed

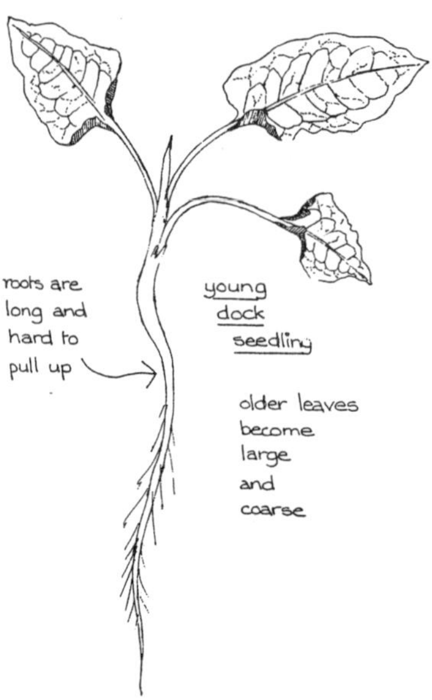

young
dock
seedling

stinging
nettle

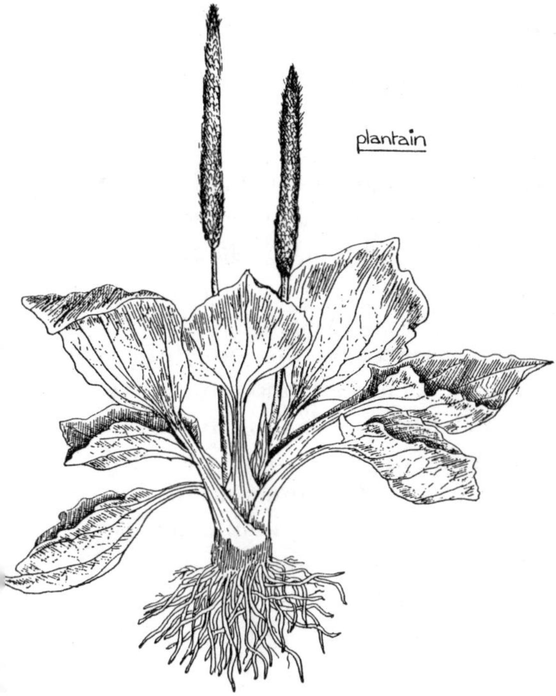
plantain

ragwort

dandelion
bright yellow flowers are carried singly on stems
fluffy seedhead
cat's ear
flower stems are normally branched
ots can
used
make
drink
at is similar
coffee
t free
caffeine
cat's ear looks very like dandelion but leaves are rounder at the tip, and less indented; they may be smooth or covered in small hairs

sow thistle

as exotics such as cherry plum, coprosma (mirror plant) and false tree lucerne.

I prefer plants with small leaves as these decay quickly, although virtually any shrub or tree is useful. Large branches are hard to use unless you have a chipper/mulcher, so stick with small branches and twigs. Eucalyptus leaves decay slowly and contain compounds that inhibit plant growth—don't use too many.

If you intend to harvest this sort of material, make sure you have the owner's permission. Not everyone appreciates bits being broken off their prize plants! Don't be a pig—take modest amounts so that the parent plant is unharmed. Don't charge in with your chainsaw and leave a desolate stump . . . unless you intend to coppice the tree.

Coppicing refers to removing the trunk so that numerous smaller branches will take its place. Not all trees can be coppiced; some just up and die. Eucalypts coppice (at least mine do) and many other trees. Coppicing can be useful for wood production, as for firewood, but I have never heard of anyone doing it specifically for mulch producton.

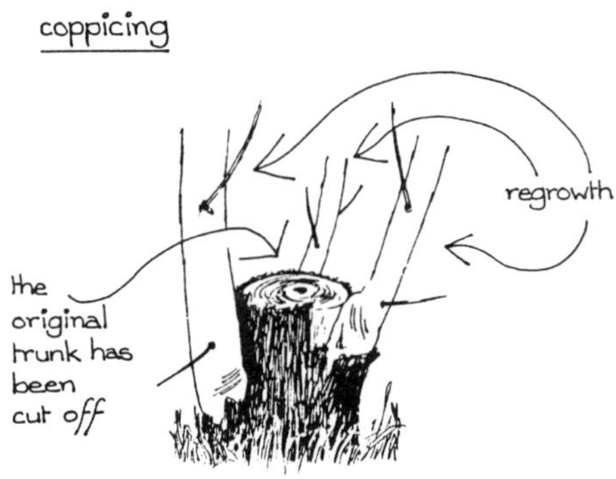

Judicious removal of branches can be seen as pruning. I do it by hand and simply break off the pieces I want. This guarantees they are not too big to be used as is for mulch. You can use a hook of some kind or a chainsaw. If, however, you gather large amounts, you have the problem of transporting awkward and possibly bulky material. A trailer or a utility would be necessary.

Your aim is to gather mulch for your garden but you should not damage your environment to achieve this.

Breaking off twigs or branches is hardly to be encouraged in the city where every single shrub and tree is doing its level best to make the world a better place. However, if shrubs and trees are plentiful and you are not greedy, the plants don't even notice you've been around.

Breakages are a natural part of a tree's life. The last week here has been incredibly windy and, as always in weather like this, branches, saplings and even mature trees have been breaking and falling in the bush around us. The sound of a large limb or tree crashing to the ground is one of life's scarier delights.

Twigs and branches are awkward to handle as mulch (unless you chip them) and you are unlikely to want many. You may well gather all you need from underneath growing plants.

GROWING MULCH

This is easier than it sounds and not as strange. It can be a small step from mowing your lawn, pulling your long grass and weeds and breaking off a few wattle branches to deciding to grow plants especially for mulch.

After all, it is this small step that led us out of the wilderness, from gathering to gardening. One small step for the gardener but a giant leap for humanity.

You can grow plants that you can cut once, plants you can cut several times, shrubs and trees and groundcovers that stay in place as permanent living mulches.

One-cut plants

These plants give of their all when you harvest them.

A time-honoured method of soil improvement, called green manuring, is when a crop is grown specifically for the purpose of cutting it and ploughing it into the ground. Alternatively, you can use these plants for mulch rather than mixing them into the soil.

Legumes are often chosen for their high nitrogen content and are sometimes mixed with cereals for *their* bulk. I frequently plant broad beans in unused beds over winter then pull the plants in spring and lay them on the ground as mulch. The pea straw left after harvesting also makes a good mulch, particularly as the next crop can be allowed to grow through it.

The roots of legumes like broad beans and peas harbour nitrogen-fixing bacteria which use the nitrogen from the air. Some of it ends up in the soil and some in the plant residue. Consequently, even while they are growing, legumes improve the soil. For more information about bacteria-fixing nitrogen in your garden, see *The Reverse Garbage Garden*.

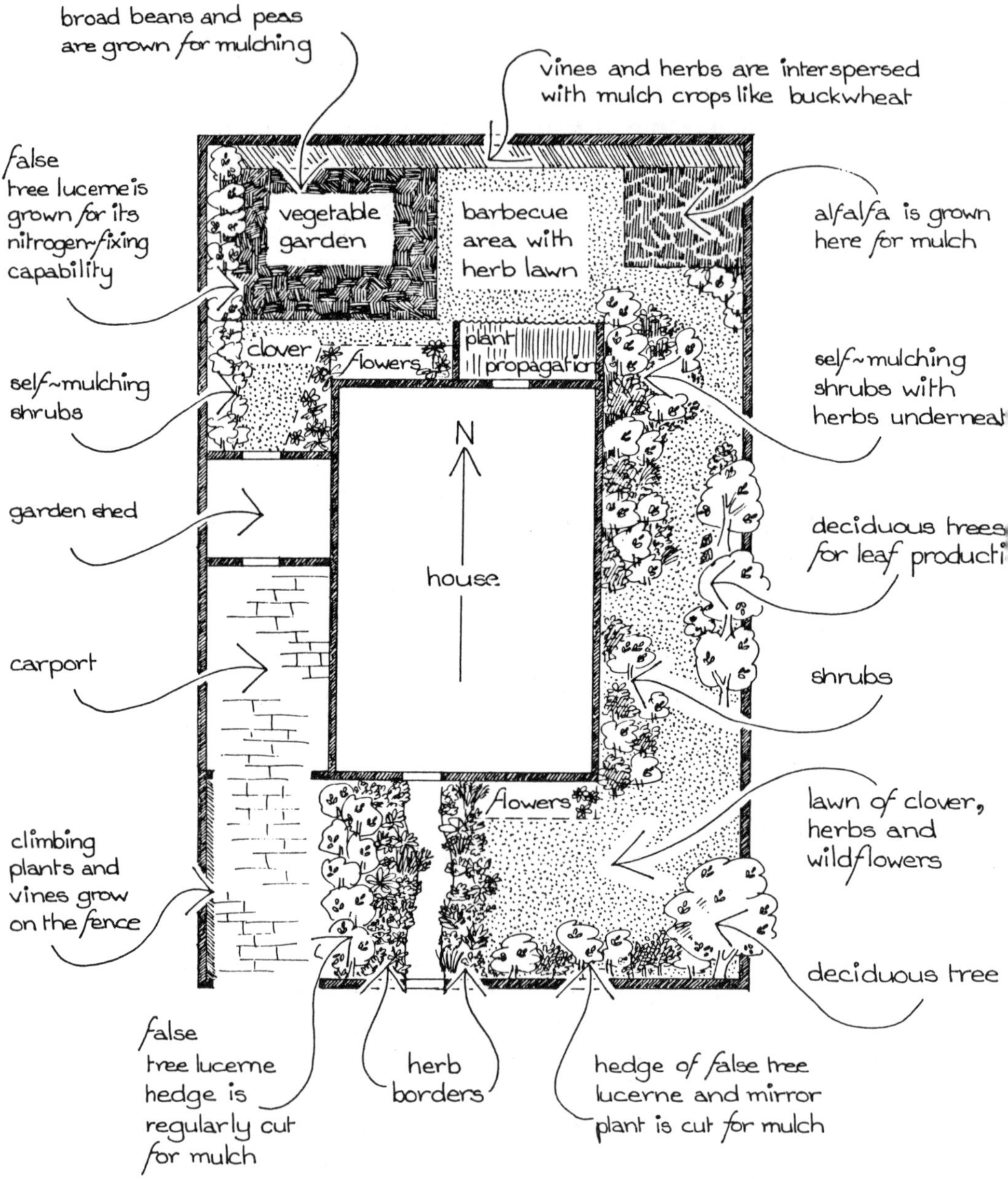

Legumes are best harvested for mulch before they crop but you'd be a fool not to go for the dual result—food *and* mulch. I *always* eat my peas and have them as mulch as well. However, it's easy to grow too many broad beans, so I harvest the plants as mulch when I need the beds for other crops.

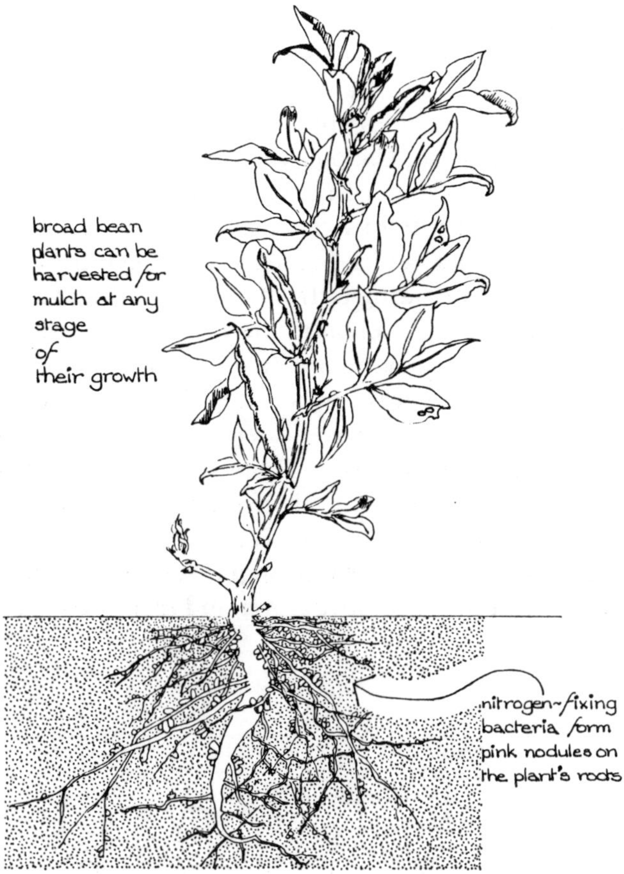

Other legumes include: chickpeas, cowpeas, fenugreek, field peas, lupins, mung beans, sesbania, soybeans, tick beans and vetches.

Cereals include: buckwheat, barley, maize, millet, oats and wheat. Sunflowers also make useful amounts of mulch.

Seeds can just be broadcast over the mulch in your garden beds, even if you have other plants growing at the time, because legumes add more than they take. Alternatively, you can pull aside the mulch and sow the seed in rows just as you would any other crop. Keep them mulched since you want your mulch plants to grow, not weeds.

I pull these plants up as I need them and lay them on the ground as mulch. They can be cut up first. If you don't pull them up, they sometimes regrow from the stumps which is not always desirable. Peas that have regrown in my garden have been vulnerable to mildew. Broad beans are useless and are prone to problems if they regrow over summer. To stop regrowth, pull the plants and mulch the area heavily.

Cut-and-come-again plants

These are plants that can be harvested more than once. Naturally, you can't just yank them out of the ground. They must be cut with blades (like scythes and sickles), brushcutters or mowers.

Lynn from Brisbane (see Chapter 10) intends to grow a few square metres of lucerne in a corner of her vegetable garden. She will cut it with a brushcutter twice a year for mulch. Lucerne is another name for alfalfa, which has a good nitrogen-fixing capability as well as providing bulky mulch. The deep ranging roots bring up minerals. As a bonus, the young leaves are edible and you can grow your own organic seeds for producing kitchen sprouts. Harvesting a few leaves won't stop the seeds maturing or the plants providing mulch, so you can eat your cake and have it.

Clover is another nitrogen-fixer. There are annual and perennial clovers, red, strawberry and white clovers, and even subterannean clovers! There are clovers for hot and cooler climates, and many grow well in shade. Most clovers are aggressive movers, spreading by seed and by creeping. Bees love clover and make excellent honey from it.

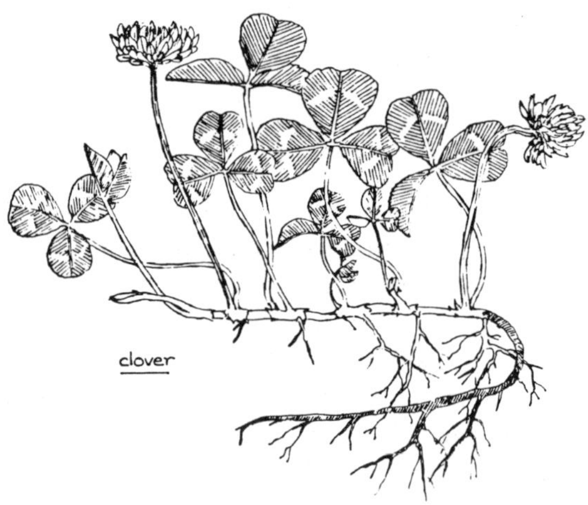

If you need advice on what clovers to grow in your area, ask farmers, locals, neighbours, nurseries and your state's agricultural department. Clovers can be mown several times a year and are both pretty and familiar.

Some other possibilities for cut-and-come-again plants are: lablab, lemon grass, pintos peanut and vetches. The first three grow in warmer areas.

Many herbs are worth growing for mulch as they provide a wide range of minerals and micronutrients. Useful herbs include: amaranth, arrowroot, borage, burdock, chicory, comfrey, horseradish, lemon balm and mints.

Herbs with pest-repellent qualities are good to add to mulches for obvious reasons. These include: marigold, nasturtium, pyrethrum, rosemary, rue, southernwood, tansy and wormwood. Many were used as strewing herbs in the Middle Ages. Now you can strew them on your garden.

Nobody is likely to grow a plant they detest just because it makes good mulch. However, mulch plants may also look good, attract bees and insects, serve as windbreaks, provide leaves, flowers or roots for kitchen use, provide fodder for animals, deter garden pests and have medicinal properties.

All the plants described here serve several purposes. When you plan and plant your garden, make each plant work as hard as possible, even if the work consists mainly of delighting you with its fragrance!

Shrubs and trees

Most gardeners like growing shrubs and trees for a variety of reasons:

1. they are perennials;
2. they take less work to maintain than annuals;
3. they can be grouped together or grown as specimens;
4. they are normally attractive and may be productive;
5. they give shade and wind shelter;
6. they help attract animals, birds, insects and reptiles;
7. they provide some of their own mulch;
8. because of their deep roots, they are less worried by soil and water deficiencies than annuals.

Now you can add another reason: they provide mulching material for the rest of your garden.

Some trees are legumes and, like the smaller nitrogen-fixing plants, return nitrogen to the soil. They can be grown in among other trees, fruit trees for example, and will actually improve the soil and fruit production. At the same time, you can harvest their branches, leaves and twigs for nitrogen-rich mulch. Leguminous trees can also be grown as a border around vegetable beds or scattered among your ornamentals.

Here is a list of some of them:

- Acacias: there are large numbers of Australian species, e.g. wattles, and exotic acacias to choose from. Try varieties that are native to your area or fulfill a specific garden need. Many are small, fast-growing and undemanding. They are good for firewood and windbreaks and are beautiful when they flower. I planted quite a few lovely small wattles several years before I discovered that they give my son hay fever—watch out for allergies.

- Albizias: good-looking, fast-growing shrubs and small trees with nice flowers that are suited to hotter areas. *Albizia julibrissin* (silk tree) grows from Queensland to Tasmania but is up to 10 metres tall.
- Black locust (*Robinia pseudoacacia*): a hardy tree with lovely flowers. It provides good honey and timber and dense shade. You can grow it from seed.
- Carob (*Ceratonia siliqua*): good shade trees, hardy with a deep tap root. Carobs will grow in alkaline soils. The pods are delicious, although they need a dry climate to ripen without getting mouldy.
- Casuarinas (she-oaks): grow in a wide variety of areas and soils. They provide firewood, pollen for bees and are good windbreaks.
- False tree lucerne, also called escabon (*Chamaecytisus proliferus,* formerly *Cytisus proliferus*): a small, fast-growing tree that *likes*

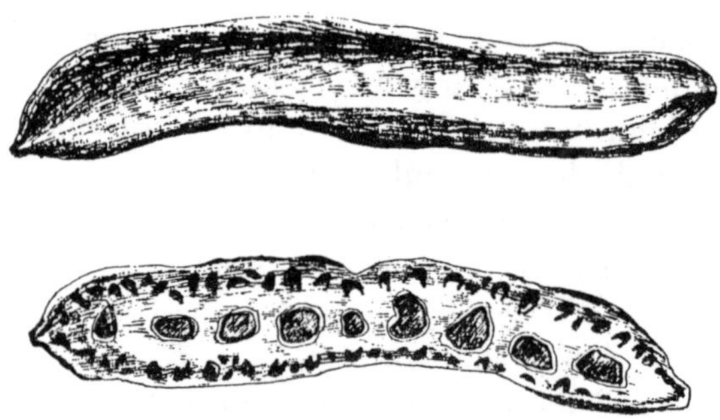

carob pods

false tree lucerne

to be pruned (that is, harvested for mulch!) and makes a good low windbreak. It doesn't mind poor soils or droughts and grows well from seed. (*C. palmensis*, also called tagasaste, is very similar.)

- Honey locust (*Gleditsia triacanthos*): good for cool climates but, as this is a prickly plant, look for the thornless variety. Otherwise, harvesting branches for mulch can be a trial! This hardy tree will grow in poor situations.
- Leucaena: fast-growing small tree for the tropics. Eden Seeds offers a small, medium or large variety.
- Pigeon pea (*Cajanus cajan*): fast-growing, small shrub which makes an effective low windbreak. The flower is yellow and red and the seed can be eaten, ground into flour or fed to chooks.
- *Sesbania grandiflora*: this fast-growing small tree is another tropical plant.
- Siberian pea tree (*Caragana arborescens*): used widely in America as a good-looking hedge, it is drought-resistant, tolerant of poor soils and frost and has yellow flowers.
- *Tipuana tipu*, (Pride of Bolivia): a South American import. This is a fast-growing, subtropical tree with yellow flowers.

There are, of course, many useful shrubs and trees which don't fix nitrogen. Anything that grows well, isn't too big, doesn't mind pruning and has leaves that decay rapidly is a candidate for a mulch tree. Here are a few suggestions:

- Bottlebrush (*Callistemon*): a decorative Australian shrub with a neat, small leaf that decays well. It attracts birds and insects and is very pretty.
- Mirror plant (*Coprosma repens*): grows semi-wild in places. This is a nice-looking shrub with glossy leaves—it isn't a fussy grower.
- Pampas grass: I know a lot of gardeners who hate pampas grass, but a relative insists that the leaves make good mulch, covering weeds well and also decaying well. Kids enjoy pulling out the feathery stalks and using them as spears and whatever. Maybe you don't want to plant pampas grass, but if it's there, use it.
- Paulownia: a new 'wonder' tree which comes to us from China where it has been grown for nearly 3,000 years. It is said to be *the* fastest-growing tree. The enormous, exotic-looking leaves are rich in nitrogen. Although Paulownias are being newly promoted at the moment, they were first introduced to Australia during the days of the Gold Rush. They provide good timber, animal feed, mulch and windbreaks, and can be grown with other plants. It's not cheap to get started with Paulownias but

they could be worth the cost. There's no need to break off branches. Paulownias are deciduous—those monstrous leaves fall off every autumn.

- *Virgilia oroboides* syn. *V. capensis*: has pretty flowers and grows well from cuttings. It is a fast grower and makes good hiding places for kids. I break off thinner branches for mulching.
- Willow (*Salix*): weeping willows are well-known, beautiful and provide canes for willow baskets as well as cubbies for kids. Cuttings root easily so you can get hold of willows cheaply. They grow along many creeks but I don't see them taking over bushland. Of course, willows tend to be large. There are super fast-growing hybrid varieties available.

All these trees can have their branches and twigs harvested for mulch, but don't forget that deciduous trees are also bountiful with their autumn leaves. We often forget the liquidambars, maples and oaks just because they are so familiar we no longer see them. If you have space, plant them.

CARPET MULCHES

Some plants can be left permanently in the ground to provide a living mulch, always in place, always working for your garden.

lemon balm

A good mix can fix nitrogen, shade the soil and provide habitats for predators (ladybirds and lizards rather than heat-sensitive aliens from outer space). Herbs and legumes go well together. The legumes (clover, for example) fix nitrogen while the herbs attract insects, provide bulk and have medicinal value for your plants. Wildflowers can be included.

This special sort of 'lawn' can be left permanently in place around your plants and occasionally cut for mulch as well. It works overtime for you. It provides a living carpet of mulch which benefits the plants around it, as well as giving you the bonus of several mowings for mulch for other parts of your garden every year.

Alfalfa and clovers make a good start. Clovers, of course, are tough enough for areas you walk on and they tend to choke out weeds. Even my sorrel gives way in a dispute with clover. Fescue is another nitrogen-fixer. It can be mixed with ryegrass for bulk. Crown vetch is hardy and will grow in sun or part shade.

Herbs could include chamomile, penny royal and thyme. Woolly yarrow is a dwarf variety with aromatic, fern-like foliage. Mints can be used, but beware—they take over. I have a friend who was in despair about controlling her mint until she started mulching with old carpet!

Comfrey has been used as a weed barrier. It's tough—no matter how often you cut or pull the leaves, they just keep growing back.

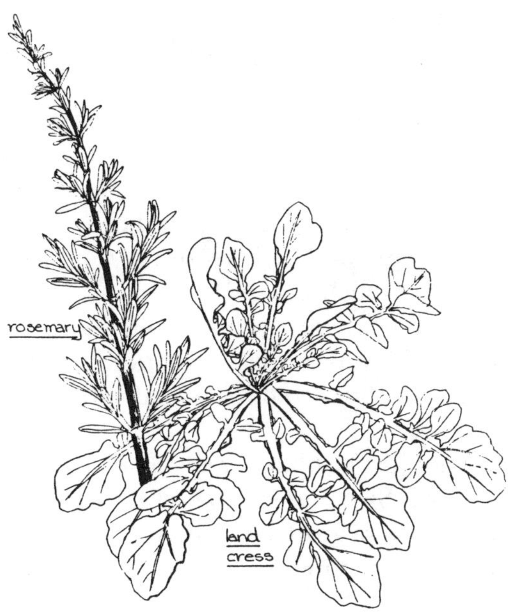

Comfrey is said to stop creeping weeds getting in, but I would be afraid that the cure was as bad as the disease! I find comfrey incredibly pushy.

Kang Kong (*Ipomea aquatica*), also called Chinese watercress, is a hardy perennial with leaves high in protein. It can be grown as a groundcover, but only in the tropics. The leaves can be eaten. Kang Kong needs good soil and water.

Alfalfa, clovers and herbs all have pretty flowers. As well, there are flowers like heart's-ease and poppies which could be added to your living mulch. They look lovely and can be naturalised since they self-sow. They attract bees and other insects. So long as they are cut or mown *after* seeding, they are self-perpetuating.

Clover, herbs and a few wildflowers—what lovelier lawn could you have winding its way around and among your berries, bulbs, flowers, fruit trees and perennials *and* you can cut it occasionally for mulch *and* know that it is fixing nitrogen year-round for you.

Some gardeners even like their living mulches to grow among the vegetables. Clover in particular, because of its excellent nitrogen-fixing capability, is often interplanted with annuals.

Mixing your living Mulches

There is no need to choose this *or* that way of growing mulch. Do it every which way. Map out your total garden areas (including lawns, paths, etc.) and plan where to put what.

If you have spots for perennials, include some mulch perennials, particularly legumes. Remember to make each plant work hard. Don't just choose a good-looker, make sure it does other jobs like windbreaking, attracting insects, providing cut flowers, food or wood.

Paths can be bordered with low shrubs or neat mulch annuals. Buildings and fences can have shrubs, trees and creepers grown against them. Vegetables can have useful borders or mulch plants strewn among them. Leguminous trees and herbs are great in orchards. Perhaps there's a place here for the troublesome but useful plants, like comfrey, dandelions and nettles.

If you have annual beds, mix in mulching annuals like alfalfa, buckwheat and broad beans or plan a rotation that includes them. Herbs can be used, too, but avoid herbs that spread like comfrey and mint.

Lawns should not just be blank green areas you walk on and moan about mowing. Switch them to useful varieties that benefit other plants while they're growing, mix in lawn herbs and flowers and put mulch borders around them. You may still be mowing, but less often and think how virtuous you'll feel about the extra high mulching value of the clippings and all this while your lawn is delighting your eye and nose!

Specimen trees can be chosen for their mulching usefulness *as well as* (not instead of) their other attributes.

Get your whole garden working for you—it will be a visual and olfactory delight and it will make *your* work much easier!

8

Equipment

The equipment you use in harvesting, collecting and distributing mulch can be as simple or as complicated as you choose.

At its most basic, all you need are your own two hands and feet. It doesn't take fancy machinery to get your kitchen wastes, grass clippings and paper onto your garden. Just carry the stuff out and put it where it's needed.

If you have commercial mulches delivered, you can spread them at your leisure. You need something to move them around in, however, just as you do if you bring home free mulches.

The smaller your garden, the simpler your equipment can be. There's no point having even a wheelbarrow if you live in an inner city terrace house with a dinky little yard out the back.

Don't assume you need this or that. Assess your needs realistically and aim for the minimum. You can always get more expensive and more complicated equipment later if you find you need it. Why waste money, time and world resources on fancy stuff that's unsuitable for *your* situation and garden?

If you are confused by machinery, don't buy anything you or a very close friend can't handle. Think about simpler ways to achieve the same end. How do you like to move *your* mulch to *your* garden?

Start mulching without relying on any specialised equipment. Learn what sorts of mulches you like to use and in what kinds of quantities. Work out where and how you can collect them. I stress the concept of what you like because what appears to be a logical solution to someone else may not suit you at all.

My husband frequently comes up with answers to my problems. He gets hurt when I ignore his neat solutions but often, very often, I don't even see the same set of problems that he does.

Personal preferences are all-important. I know several older women who have chipper/mulchers and love them. They pay to have them serviced and fixed when necessary. The machines save a lot of bending, collecting and fussing. When these people were thirty years younger, they probably wouldn't even have wanted them.

Assess what *you* and *your* garden need and be prepared to look at the whole question again any time your circumstances change.

Don't get in over your head. Start small and you can always move up a notch.

HARVESTING MULCH

Many mulches are gathered rather than harvested. You don't have to cut the seaweed on the beach, only parcel it up. It is the same with animal bedding, cut grass, leaves, manure, paper and pine needles.

However, weeds have to be pulled or cut, grass mown and branches harvested. There are many tools, either hand powered or motorised, that can help.

Hand cutters

Your unaided hands are useful (protected by gloves if you prefer), though not over large areas. Hand pulling of weeds and long grass may work within your garden and around its edges. I doubt if anyone can manage more than this.

I break branches by hand from suitable trees for mulch. I use bottlebrushes, daisy bushes, false tree lucerne, virgilias and wattles. If the branch is too large to be removed by hand, it will be difficult to use as mulch.

There's a range of tools that consist basically of a wooden handle with a blade attached. The scythe was used for centuries to harvest grains and hay. There are also bracken hooks, fern hooks, sickles and many others with curving blades and similar names.

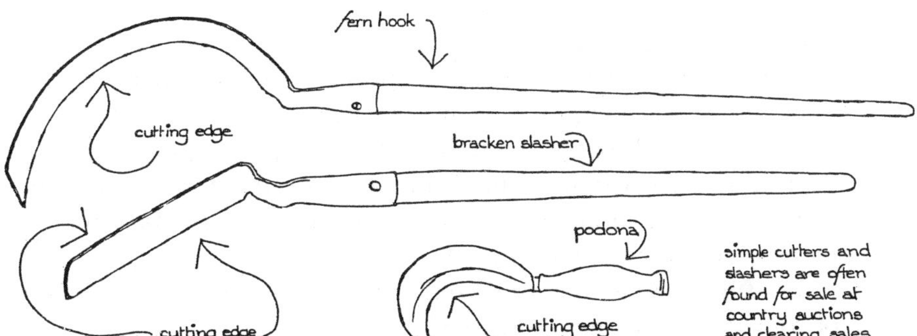

Many of these are fine tools for smallish areas. All are hard work to use and tough on your back. It's easier if you keep them very sharp. Larger areas can be tackled if you break the job up and spread it over weeks.

Scythes are different. The handle is very long so that bending is minimised. It is possible to scythe large areas—that's how grain crops were harvested in the Middle Ages—but in practice scything is difficult, tiring and time-consuming. Using a scythe is all about technique rather than brute strength and not many of us have the technique. A scythe has to be razor sharp. Scythers used to spend one hour working and five minutes stroking the blade with a whetstone.

I have used a scythe for grass control in the past. Once you get a rhythm going, it's okay, but easy it's not.

Tagasaste and escabon are grown in the Canary Islands for animal feed and are harvested using a small curved knife called a podona. These are marketed in Australia. I have never tried one but they would be less damaging than breaking the branches off trees as I do, and faster and more practical than secateurs or a handsaw. It is hardly worth starting a chainsaw to cut a few branches for mulch!

These are the simplest tools I know of. I discount scissors as a method of cutting grass even if a friend did do it as a kid! Hand tools don't seem practical to me. Most of us have to fit our gardening into a busy and varied lifestyle. We just don't have the time for such labour-intensive methods.

Moving into greater mechanical complication, there is the lawn-mower that works with you as its engine. The ones I've used have been so old, blunt and rusty that they weren't much fun. It is very difficult to keep the blades sharp. Hand mowers *do* work, but use a machine in good condition and don't try if the grass is knee-high. In this case, use a brushcutter to get it down first, then the mower to keep it short.

Mechanical cutters

For cutting grass there is the basic option of a brushcutter or a lawn-mower, although they don't do quite the same job.

Lawn-mowers are best with grass that is mown regularly and kept short. They can be difficult to use on steep land or even mild slopes. There are attachments that go behind tractors, ride-on mowers and mowers you propel.

The tractor attachments are only suitable for very large areas and, of course, you first need a tractor!

Ride-on mowers are best for large expanses—they are expensive, they can be dangerous on slopes and they can't manoeuvre in tight corners. It makes you feel pretty powerful to sit there while a machine does the work. Sometimes I wonder if this is why people buy them. I guess it keeps you up above the snakes.

Ordinary rotary mowers, as used by most people, are an Australian development. They are effective and brutally simple. In most, the motor sits above the blade and rotates it via a shaft. In some, the motor is behind the blade and rotates it through a belt. There are also self-propelled mowers—you walk behind them and they cut up anything. The Mobilco heavycut is a large tough version, good on hills; a go-anywhere machine much used by farmers for weed control.

I'd like a rotary mower. They can be hard to start and are anything but maintenance-free, but constitute a fairly painless way to cut large areas. They use more petrol than brushcutters but less human energy. Most are difficult on slopes.

Make sure your mower has an attachment for catching grass. I have had many a fight (that is, amicable discussion) with my husband about this. He says it is too much messing around to stop and empty the catcher. He *claims* it's easier to rake the grass afterwards. However, *he* doesn't do the raking, I do. It's not easier. Use a catcher. (My husband continues unrepentant.)

Brushcutters are terrific inventions, very like motorised scythes—the early models were called auto-scythes. I love cutting a swathe through high grass, leaving long windrows ready for gathering. They can be used on slopes and long grass. They even cut through blackberries, bushes and saplings. They are not so good on lawns—too much effort for too little result.

The harness that comes with a brushcutter stops you chopping your foot open, but they are potentially dangerous. The cutting blade is horribly exposed although some have rudimentary casing around it. Watch out for cats, dogs and kids. I once chopped up a snake without knowing it was there. That wasn't so bad, but look carefully if you have a deaf cat that likes sleeping in the grass.

As well as blades, brushcutters can be fitted with a fishing line that whirls around (whipper snippers have this). It's less dangerous but not so good on long grass. They can also have saw blades and are used in America and Norway to slice effortlessly through 3 metre high Christmas trees.

Brushcutters are incredibly noisy; rather like chainsaws. They are heavy to use (for me, anyway) and vibrate like anything. Leaving bits of wood, bricks or glass jars lying around can wreck the blade. Make sure you wear a helmet and goggles as well as earmuffs.

The ideal for a large place—.2 hectare or more—is a lawn-mower (with catcher) for lawn areas and a brushcutter for wilder areas. A good compromise is to own a mower and to hire a brushcutter from time to time. The ideal for a small place, such as in the inner city, is to have neither. Avoid grass altogether. Design a garden with pathways, a paved barbecue area and gardens.

For in-between gardens, you'll probably need a lawn-mower (motor or hand) or easy access to one. Although the idea of sharing one between a few friends or neighbours is attractive, a lot of people rightly warn against it. Sharing machinery is fraught with trouble. It may be better to preserve your friendships and buy a mower, or landscape your lawns away.

My husband insists that two-stroke engines have come a long way, but I still think they can be difficult to start and they need regular maintenance.

If you have plenty of time and you enjoy challenges, try a scythe.

You can, or course, use these mechanical devices to cut other people's grass and lawns. I don't advise it, unless you're getting paid. If you just want other people's waste as free mulch, wait until they cut it themselves.

GATHERING YOUR MULCH

There's your hands again. Grass and leaves can be shovelled into bags, buckets or billycarts (wheelbarrows if you don't have billycarts). I'm even prepared to harvest *some* manures by hand, namely not-too-fresh horse manure, but I admit my girls think this is disgusting.

If you want to keep your hands clean and your nails unbroken, you can always wear gloves. On the other hand, if you're *that* fussy, maybe you should forget gardening and take up embroidery.

Hand tools

Rakes don't fit neatly into a category but are very useful for raking cut grass and lawn clippings into usable piles. When I had more time for collecting mulch, I used to carry a rake in my car so I could gather other people's scattered grass. Now I demand that it already be in piles.

Garden forks, pitchforks, shovels and spades can be used to grab hold of your material. It can then be put into any suitably sized receptacle, from a plastic carry bag or a bucket through wheelbarrows and garbage bags to trailers and the backs of utilities.

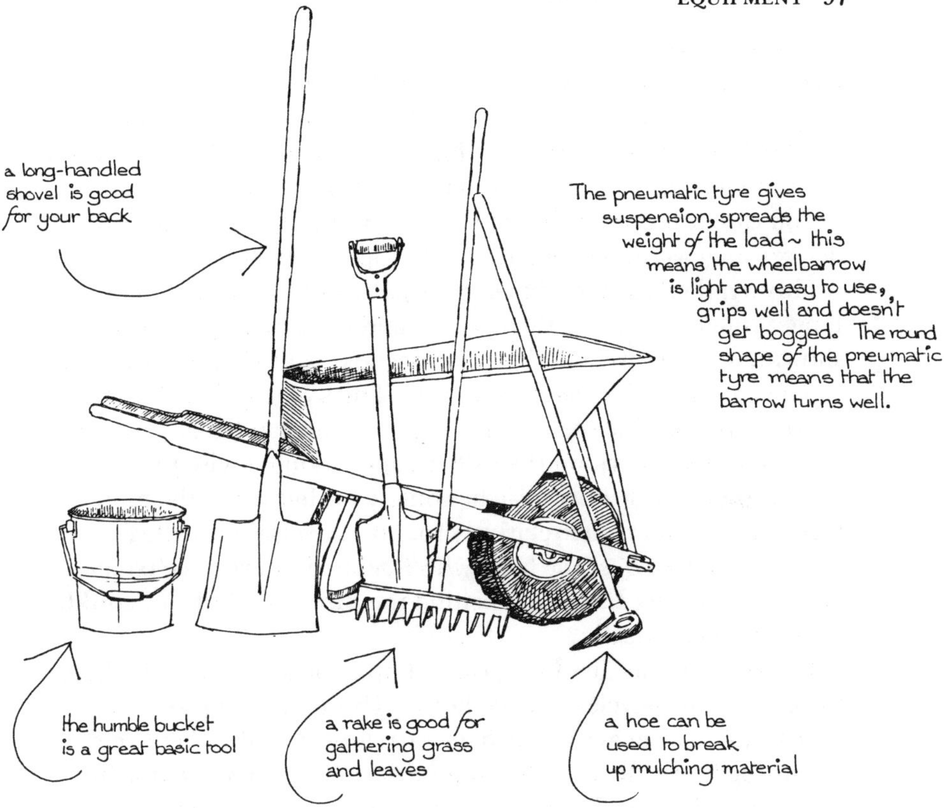

Shovels and spades are good for manure, while garden forks or pitchforks are excellent for material that sticks together in slabs like hay and straw, or animal bedding based on hay. I find dry leaves most easily pushed around by hand as they are flighty creatures. Wet leaves can be manoeuvred with a spade.

If your lawn-mower has a catcher, grass can be emptied directly onto the garden or into a container.

Although I know there are mechanical devices for sucking up leaves, I am not aware of anyone who uses them for collecting mulch, so I'll go straight on to . . .

CARTING YOUR MULCH

This covers getting your mulch onto your garden. You may be transporting the material a few metres from the edge of your garden beds or a few hundred kilometres from the beach or a city park with its fantastic accumulations of autumn leaves.

Hand carting

Occasionally I only have to carry mulch from one part of my property to another. The longest distance this involves is a few hundred metres. Sometimes I simply push leaf mulch into a couple of plastic buckets and walk it to my garden.

You can also use cardboard boxes and carry bags or wheeled devices such as prams, pushers and shopping trolleys. These are readily available at Op Shops and tips or from friends who no longer intend to extend their family and whose equipment is showing distinct wear and tear (and whose isn't after a few kids?).

There is also movement by bicycle. This is leg rather than hand power. The humble basket hanging from the handlebars can be used, or a small pack on the rider's back. I don't advise putting messy, smelly or dribbly mulches onto your back! For the more sophisticated, there are relatively roomy contraptions called panniers that go over the back wheel. Sometimes you see paper deliverers pulling little carts behind their bicycles, but this could be a hazard in city traffic.

In Vietnam, the Viet Cong carried up to 90 kilograms of whatever by using their bicycles as pack horses. Grain bags can be ingeniously slung across the back carrier or, like the Viet Cong, across the frame. Reasonable amounts can be moved by attaching something like a milk crate (I said 'something like' because, of course, keeping milk crates is theft) to the back carrier.

In Melbourne recently I saw a loaded bicycle that wouldn't have shamed even a hardened Viet Cong traveller. Plastic bags were slung over the frame front, back and middle. Bags were hanging from other bags. The whole thing was higher than the owner (who was walking!) and the bicycle could hardly be seen.

Don't forget the noble wheelbarrow. Don't bother with one unless your garden and property size warrant it but, if they do, it is the mulcher's Number One tool. A wheelbarrow saves me an awful lot of walking around with buckets. Wheelbarrows are good for all materials and are easily loaded using forks, shovels and spades.

Even wheelbarrows have undergone improvements. For years I insisted we didn't need one of the fancy wheelbarrows with a pneumatic tyre. No—my little, old, rusty, jerky, bumpy, metal wheelbarrow was just fine, thank you. Am I glad my husband finally won this argument! Nothing, *nothing*, would make me go back to a wheelbarrow with a solid tyre—pneumatic tyres make it so much easier to use!

Wheelbarrows are one example of synthetic being superior. Metal wheelbarrows rust out their trays. The plastic (polycarbonate) ones are tough, don't appear to decay in sunlight and don't rust when you leave them out in the rain. Somebody, somewhere, said the wheelbarrow is one of humanity's three great inventions. Don't ask me what the other two are.

You can also shift material like grass, hay, leaves and straw by using a large square of tough material, like a tarpaulin. Any old tarpaulin will do. You can sew handles into the corners to make it easier to use. These are advertised in American gardening magazines, and may be available here. You can pile them up with material and drag them along the ground, either opened out or with one end pulled up and covering the mulch. If your tarpaulin is not too big, you can fold the edges together over the material and carry it on your back.

You can't drag them on slopes and over very rough ground as things like tree stumps and branches tend to tear them, but they are good for carrying material in places too rough or steep for wheeled traffic.

Mechanical carting

For carting material a car is the most obvious machine to use. Bags and boxes of mulch can be stuck in the boot, strapped to the roof rack, loaded on the back seat, perched between passengers or at their feet, and put in the open back of a station wagon. I have done all these things. In an emergency I have even loaded cut grass straight into the back of my station wagon, but you have to be crazy to do this. If you own a motorbike with a sidecar, don't feel shy—go ahead and stick a few bags of leaves in it.

Be careful with seaweed. The salt isn't exactly good for your vehicle.

Utilities and station wagons are more useful than sedans. A trailer keeps potential spills and mess out of the car but is only economic if you regularly need large quantities of mulch. You can hire trailers, which has the advantage that the seaweed will rust out somebody else's equipment!

Cars can be used for small to medium quantities of mulch. If you keep bags in the car, you can pick up mulch wherever you happen to find it: cut grass on the side of the road when you go shopping; leaves at a friend's place where you've stopped for a chat; seaweed when you take the kids to the beach. You can bring back one bag of mulch or ten—it's up to you.

Animal feed bags, briquette bags and hessian bags are all good. At my tip, I can pour my rubbish into a sort of hopper and reuse my plastic rubbish bags. They're only good for one or two extra jobs because of their flimsiness. Don't try to reuse bags with holes in them; it doesn't work.

Most feed bags now are synthetic white ones (polybags)—they come apart if they get torn and quickly become brittle in sunlight. Getting wet doesn't hurt them, though. Farmers, hobby farmers, friends with a milking goat or backyard chooks or dog owners who buy dog biscuits in bulk can all be a source of bags.

Tractors can be used to shift mulch within a property or short distances along the road. If you live in the city, forget the tractor, try the bicycle.

SPREADING MULCH

Spreading involves the same kind of tools as hand gathering. Mulch has to be removed from the vehicle it came in, taken to the garden and spread.

If the material is bagged, it is easy to handle. Bags can be dragged or taken by wheelbarrow. Loose material can be removed using hands, garden forks, pitchforks, shovels or spades and put into bags, buckets, wheelbarrows or other wheeled devices (e.g. prams and pushers) or onto a strong square of material which is then pulled to the right spot.

If the mulch is being spread over a large area, it can be thrown onto it with a fork or shovelled into place. To spread it lovingly around seeds, seedlings or favourite plants, you can't beat the hands. You can then make sure that the weeds are covered and the plants are not.

Bending is one of the activities that makes gardening tough on your back. Using a long-handled shovel rather than a spade helps. Choose pitchforks with long handles as well.

STORING MULCH

Mulch is best stored under cover when possible. Rain leaches the nutrients out. If it can't be covered, keep it where the nutrients will benefit plants, such as parked under fruit trees.

Don't keep paper out in the open. If it is dry it will blow around; if it is wet it will fall to pieces. Sheets of plastic are good for covering heaps of mulch, as are bags, corrugated iron and wood.

Ideally, excess mulch should either be composted or kept in a shed or under a roof. Don't store combustible material like

Shredded bark being spread around Australian shrubs.

Bonnie is putting cut long grass around her apple tree.

Sawdust and grass are helping to grow tasty strawberries.

Corn and peas are growing in well-rotted mulch.

Seedlings are a great by-product of using mixed mulches on your garden.

These abalone shells will add valuable nutrients, but we'll cover them with hay as soon as we finish taking this photo!

Clover can be cut to provide a high nitrogen mulch.

Herbs can be cut and used as an aromatic and pest-repelling mulch.

This dense stand of peas has quickly become self-mulching.

Sawdust can be used to protect seedlings from slugs and snails.

Wood ashes are protecting these lettuce seedlings from slugs and snails.

Slugs and snails cluster under slabs of mulch and can be picked off and killed.

Leaves and grass keep tomatoes off the ground.

Fresh yellow hay surrounds a healthy rainbow chard plant and several potato plants.

cardboard, dry grass, hay, paper and pine needles against buildings or under your verandah.

Mulch should not have to be stored for long since it achieves its garden purposes best when it is constantly renewed. If you temporarily have too much, you should soon be able to use it.

CHOPPING UP MATERIAL

Hand choppers

Many gardeners tear paper up by hand and then wet it so that it decays faster. Apart from that, you must use tools.

Kitchen rubbish can be hacked around with a hoe or the end of a spade. It's worth doing this either for compost or mulch. Since you're normally dealing with very small amounts, the work is not as hard as it sounds.

Otherwise, it's hard to break up materials without using machinery.

Mechanical choppers

You can, theoretically, use a rotary mower to shred material, but it's not always successful. You put stuff like seaweed on the ground and run over it—it is slow and you have to keep raking the material up so the blades can reach it.

You can also modify a rotary mower by making a hole in the upper casing and building a hopper on to it to feed material into the blades. This works with light material. It's definitely not my scene, so I'll simply float this as a possibility and not try to get technical.

Chipper/shredders

I can't see any reason to own one of these unless you have at least a suburban-sized block. You can buy chipped material from many city councils and some chip ratepayers' prunings free once a year. You can also buy chips from private tree-loppers.

Even in the suburbs and country towns a machine like this is hardly worthwhile unless you have access to reasonable quantities of branches and leaves. Perhaps you have a country property or a friend's farm to collect material from? Perhaps you know your friends and neighbours are dying to donate their prunings to you?

It is also possible to get together with others and purchase a machine between you. This makes good economic sense and the

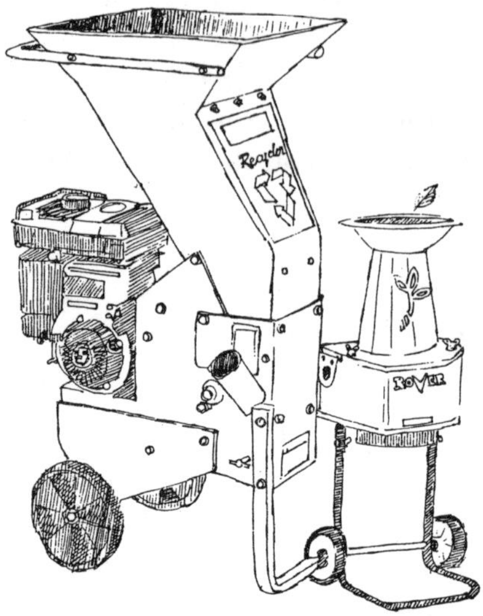

chipper~mulchers

machine will spend less time being idle. However, sharing machinery can cause a lot of friction.

Another option is to buy a solid machine for your own use (not one of the smaller models) and to recoup the cost by hiring it out to other gardeners. Work out exactly how you intend to advertise your service. Make enquiries at places where such hiring is done to find out what to charge and what sort of deposit to ask for. Expect your machine to take a beating.

Okay, so you've decided to get a chipper/shredder. They're terrific if you have a lot of branches, kitchen waste, leaves, plants and twigs. They are *not* necessary and are quite expensive, but they are useful. There are electric models and petrol-driven ones.

All the units I looked at worked in much the same way. There was a hopper where you fed the material in. This led to a chamber where blades cut material and flung it against screens. The pieces went through the screens when they were small enough. On many you are able to attach a bag for collecting the chipped material. Otherwise, you can put a tarpaulin under the opening. There may or may not be separate openings for branches and for lighter material.

Electric chipper/shredders

Generally, electric-driven machines are smaller and lighter. Rover, an Australian company manufacturing its products in Brisbane, has two electric models. The smallest, the Mulcher, takes branches up to 20 mm thick. It's fairly slow to use and not suitable for large quantities. It is small and manoeuvrable, but I consider it too small to recommend.

The Magnum Chipper Chopper made by Allpower Industries is similar. It is light and easily moved and has a separate opening for feeding in branches.

Rover's next model up, the Maxi Muncher, is still light and easy to handle but takes branches up to 32 mm in diameter. I wouldn't recommend going smaller. I know several older gardeners with units similar to this one who are very happy with them although they are a little slow to use.

Masport has a larger electric model which takes branches up to 65 mm thick. If you want an electric chipper, this could be worth checking out. At about $900, it costs much the same as the smaller petrol versions.

The smaller electric models retail in the $400s and the larger Rover is around $700. I found variations in price of more than $50 per unit, so shop around before buying.

Petrol-engined chipper/shredders

Rover also sells a larger petrol-driven mulcher, the Shredder, for nearly $1,000. Masport and Bolens both have several petrol-driven versions and Motorcolt sells Italian models that look very good (and cost accordingly). Differences are in the size of the engines, from 5 hp up to 7.5 hp. Naturally, the larger versions are sturdier, take larger material, work faster and cost a lot more. They are also harder to manoeuvre around.

Fauser Worms in South Australia sells a shredder which is the same as the one they use to produce their worm food. It has a separate hopper for branches and looks sturdy and well-balanced.

The larger models seem to have separate chipping blades that chip branches while the hammers munch up the material that is fed in through the large hopper at the top of the machine.

Bolens markets a machine called a Chipper Vacuum which sucks up garden rubbish and shreds it. It also has a hopper for feeding material in. I guess if leaves on the 2 hectare lawn are your problem, this could be the answer.

For very large jobs, there are units that fit onto a tractor—a great option if you happen to have a tractor!

Chipper/shredders with petrol motors cost anything from $1000 to $3000. For the higher price, you get a fast worker and a tough unit you could hire out. It's a lot of money, so check that's what you really need.

An alternative to buying is to hire a machine. Last time I enquired it would have cost me $60 for one day. You can do a lot in a day, especially if several gardeners combine and you make sure you have a stack of material ready. When you hire, maintenance problems belong to somebody else!

Using your chipper/shredder

Don't put your hands down into the hopper. Treat the blades with respect at all times. Keep them sharp but don't go near them when they're in use.

Don't overload the hopper or feed material in too fast—this causes the engine to stall. Soft material like garbage, manure and wet leaves tend to clog the machine. Alternating soft and hard materials helps—wood chips, for example, shift the soft material.

It's a good idea to use prunings quickly. Soft, green material is easier to shred than hard, dry stuff. Chipping the hard, dry stuff shakes the machine around. Run on softer material, your chipper/mulcher is likely to be easy to maintain.

Check the oil regularly, keep the blades sharpened and the shredding chamber cleaned out. Or get a friend or partner to do this.

I have mulched heavily for a decade without using a chipper/shredder, but boy would I love to have one!

HOW MUCH EQUIPMENT DO YOU NEED?

Look, you don't need anything beyond a pair of willing arms and legs and a few bags and buckets. Moving up the scale, you get to shovels, spades and wheelbarrows. You can hardly include the car, since nobody buys one specifically for collecting mulch!

Brushcutter and lawn-mower come next—they're not necessary but many of us need them anyway to keep the grass down. As a bonus, they give us excellent mulch.

At a more sophisticated level, you might like to chip and shred your own leafy, branchy, rubbishy mulch. Nice work if you can get it.

Remember to keep realistically assessing *your* needs, desires and expectations. Never get equipment to keep up with the Joneses or because you think you ought to. Make sure you need it and want it.

9

Everything You Ever Wanted to Know About Mulch

All right, I admit it—the title of this chapter is a swindle. I can only tell you 99.9 per cent of what you want to know.

I've had a lot of questions put to me over the years, ever since I've been visible collecting mulch, listened to talking about mulch and more recently caught out writing about the stuff. I have tried to remember the more common questions and (hopefully) supply answers.

I've even made up questions I think you *might* ask or *should* ask. I tried to get my husband to think of some, but 'is it stable in a choppy sea?' suggests his mind was on kayaks not compost, sailing ships not seaweed.

Some of these questions have been covered in previous chapters. However, I feel it is useful to bring them together here because I keep getting the same kinds of queries over and over, even from gardeners who already use mulch.

Can I use mulch everywhere?
Sure.

Don't let your plants get overwhelmed by the mulch. This means fine mulches around seedlings, coarser mulches around larger plants. Mulch keeps soil cool in spring and summer so, if you need a warm soil, pull the mulch off that section of the garden to allow the sun through—but make sure you put it back on later.

Define what your aims in using mulch are, and what your garden needs. Is the mulch mainly for weed control or is soil fertility as important or more important? Choose your mulches accordingly.

All plants appreciate the benefits mulch brings.

How often do I have to put it on?

There are three main aims with mulch: conserving soil moisture; improving the soil; and getting rid of the weeds. You have to put enough mulch on to stop weeds growing up from the ground. You must then keep putting mulch on. This is to prevent vigorous weeds from making it through the mulch, to stop weed seeds on and in the mulch from germinating and to knock such seeds as *do* germinate on the head before they get big.

There is no magic rule about when to put mulch on. Mulching is a forever kind of job, not something you do in spring and then forget for a year. If mulch becomes available, *that* is the right time to apply it. A sure-fire indication that you need mulch is the appearance of weeds. As soon as you start finding weeds, get more mulch.

If you have an underlay of cardboard, paper, plastic or similar materials and something seeds don't like on top, e.g. sawdust or wood chips, you shouldn't need to mulch for a year after setting up your garden. If you mulch with seedy material like hay, you'll probably need to start looking for more mulch as soon as the hay is put on.

Just keep covering those weeds and you can't go wrong.

When do I stop putting on mulch?

Hard to say. The basic answer is never. You are going to go on using mulch for as long as you can make it out to your garden.

However, you don't want to swamp your plants. What if you have a very small garden (this is the famous what-if game often played obsessively by kids) and what if you are a super efficient collector of all sorts of mulches? You could end up, temporarily, with too much mulch. You can store it, compost it or give it away to friends.

Soon enough, the mulch in your garden will have begun the fast slide into decadence and you'll be putting it on again.

Just keep putting on mulch unless your plants start disappearing under the weight. Then you can take a break for a week or two.

Why does my mulch keep disappearing?

I hate to be the one to tell you, but that's what mulch does!

Apart from inorganic materials like plastic, mulches decay. Some, like wood chips, decay very slowly and some, like hay and manure, decay quite quickly. As mulches decay they add organic matter to your soil and nutrients are released that aid plant growth. Worms pull pieces of mulch down into the soil and all sorts of visible and invisible organisms break the material down.

This is one of the purposes of mulch.

As it decays, the mulch disappears. This is one of the reasons you need to make a commitment to keep on mulching if you want your garden to thrive.

If you want mulch that doesn't disappear and you're sure your plants won't suffer (that is, if soil improvement is not your major aim), use one of the long-lasting mulches. Stones might keep you happy.

What if it doesn't rot?

Mulches decay at different rates under different conditions. If it is kept dry, for example, paper doesn't rot much. If it is wet, it can disappear quickly.

If your mulch rots rapidly, you'll know you are supplying plenty of organic matter and, hopefully, plenty of nutrients. If your mulch is slow to decay, it's probably less use as a soil conditioner and improver.

Define your aims. If weed control is your main aim, don't worry about slow decaying mulch. It will do its job longer that way.

If soil improvement is important, make sure you include mulches that decay rapidly. This generally means including material that is high in nitrogen, such as animal bedding, grass clippings and manure. If you have stubborn mulch which is slow to decay, throwing high nitrogen material on top normally helps.

Remember, slow decay doesn't matter unless you want fast soil improvement.

How do you find enough mulch?

This only looks like a problem when you haven't tried it. It is amazing how, once you start looking for mulch, it's suddenly all around you. It's like when you buy a car, say a Fiat. Suddenly you're seeing Fiats in the street, in the car park, on your way to work and outside your dentist's. The Fiats were always there, you just didn't notice them.

We are all walking, driving, jogging past mulch all the time. If not this week, then next. All we have to do is notice it.

For greater detail about finding mulch, have a look at Chapter 3.

Commercial mulches, of course, are there for the asking. All you have to do is work out how much you want and how it is to be delivered.

Oh—and pay for it.

I've just cut my grass and there's too much to use; what can I do? Nonsense. There is never too much to use.

This question was put to me by a friend. I advised him to mulch under his trees with the grass. He said, if he did the trees would be smothered, covered to their crowns in grass. I said no way.

So he tried it. He put the grass 30 centimetres deep around a couple of trees and he didn't even manage to mulch right out to the drip line!

Mulch never seems to go as far as you think it will. If your vegetable garden is fully mulched, mulch your fruit trees. If these

too much mulch, perhaps?

are okay, mulch your ornamental plants. By then it will be time to start with your vegetables all over again.

If you *do* manage to accumulate, beg, borrow or steal more mulch than you can use, store it in some convenient spot until you're ready for it. If it rots a bit in the meantime, so what? You can also compost excess material to make an excellent soil conditioner. For information on making compost, see *The Reverse Garbage Garden*, Chapter 8.

Isn't it too much work?

No! No! No!

Mulching replaces hard work with easy work. Sure, it involves work—what doesn't?—but not a lot.

Digging and weeding can go. Instead you look out for mulch (not work, just a matter of developing the right attitude), collect it, bring it home, get it to the garden and spread it.

These are slow, easy jobs and I spend very little time on them. Most of my work is in lifting rubbish bags of grass into the car and out again at home. It's up to me when I put the stuff around my plants. I like to do it promptly so the grass is nice and fresh and I'm ready for the next lot. However, it will keep.

Most mulches can be stored somewhere convenient and used when you have a spare half-hour.

The other work that mulching saves you is fertilising and watering. If you pull weeds, you're not improving your soil fertility. You have to put *something* into your garden to keep the plants happy. Most of the mulches you use supply organic matter for the soil plus some nutrients.

In some gardens, with some plants, mulch eliminates watering completely. No standing around with hoses, turning taps or moving pipes. Even if you still need to water, you will need to do so a lot less. Less watering = less work.

I spend no more than half a day a week, on average, working in my garden. For that time, I get a whole year's supply of vegetables for my family of five, about half our fruit, raspberries to jam, freeze and sell plus plenty of flowers. Most people think I spend all day in the garden. I would have to—if I didn't take the easy way out and use mulch!

Doesn't it look messy?

Quite the opposite. A mulched garden looks neat, tidy and productive.

Bare soil looks terrible to me. I can't help thinking of the sun baking it, the rain beating down on it and the weeds laughing at it. Weedy gardens are okay by me, but tidy? No way.

Mulch covers the bare soil and stops the weediness. Some materials are inherently messy, like orange peel and eggshell. Manure is in-between, not quite messy, not quite neat.

No problem. Simply cover these mulches with tidy mulches. Grass clippings, hay, seaweed, stones, straw, wood chips—all these and more are not just neat, they add colour, depth, variety and warmth to your garden.

Japanese minimalist gardens use stones for their austerity and serenity. Nobody assumes the stones are messy just because they stop weeds coming up. Most mulch looks terrific—it's perfectly natural, after all, for the ground to be covered in litter. That's how soils are built up.

If you imagine your garden as a three-dimensional, living changing work of art, the mulch is a vibrant element, a rich pigment to use.

Isn't it better to make compost?

This is a matter of opinion. Some gardeners swear by compost and most organic gardeners make some compost.

I prefer to put my material straight on the garden. It's easier and simpler than composting and means that your garden benefits immediately. Although nutrients take time to filter down, mulch protects the soil from rain and sun, and harbours insects, microbes and soil life. Nutrients *are* slowly washed out to your plants' benefit.

When materials are put into a compost heap, they can't help the soil and your plants until the compost is finished and applied.

It's wrong to see this as an either-or situation. Do both—make compost *and* use mulch.

Won't it stop rain getting through?

Materials like grass clippings and pine needles can mat together and stop rain reaching the soil. The solution is not to put large quantities on at one time. Lay these mulches down in thin layers. Leave them for a few weeks before putting more mulch on top.

Alternatively, mulches that might mat can be mixed with other materials like hay, sawdust, straw or weeds. Thick layers can then be put on the garden.

Some mulches are impermeable—corrugated iron and plastic, for example. You can punch holes in the plastic, but it rather defeats the purpose of using it since weeds will grow through the holes.

Cardboard and paper can also stop rain getting through unless they are kept thoroughly wet so that they rot quickly.

Use impermeable mulches only where it doesn't matter that water can't get through. In commercial strawberry farms, soaker hoses are often laid under the plastic mulch.

New weed mats (plastics) for garden use are permeable, so these can be used when a porous but efficient mulch is needed.

Can you really use yukky things like banana peel?

Sure. Yukky things like banana peel, orange skins, the outside leaves of cabbages and those mouldy carrots you discovered in the back of your fridge can all go straight on your garden so long as you cover them with some nice clean mulch like grass, hay or straw. The mulch on top keeps out flies.

These sorts of materials have a lot of terrific minerals to contribute to your garden plus they provide organic matter. Worms love your kitchen rubbish, too.

If you are in a situation where it is impossible to put kitchen rubbish straight on the garden (because of dogs, neighbours, regulations or whatever), you can dig it into the beds or compost it. Avoid putting meat scraps on top of the soil—compost them, dig them in or give them to the chooks.

It's better for you to find a way of usefully disposing of your own kitchen wastes rather than shoving the whole problem onto your council.

Will it smell?

Mulch isn't unpleasantly smelly. Wood chips, surprise surprise, smell mildly woody. Grass clippings smell grassy. Mulches, being harvested and dead, lose their individual smells fairly quickly, just as their colours fade.

If you put a pile of putrefying material down in a lump it smells rotten—but why would you do it? Especially in your own front or backyard. A compressed and wet pile of grass clippings can smell horrible, but a thin layer of clippings around your plants doesn't. Dubious material should be buried under clean, tidy materials—kitchen wastes, for example, under hay.

Seaweed has a relatively strong seaweedy smell, although it, too, quickly fades. If you don't like the smell, don't use the material. There's enough to choose from, after all.

The 'designer' mulches, the commercial ones, certainly don't smell. Put on properly in thin layers with messier material covered, no mulch smells.

Can I bring in plant diseases and weeds?
Yes.

There is risk in everything we do. Only *you* can decide what's right for you. The only problem I have ever had was bringing in couch grass and that was my own fault. I knew it was couch grass, but I still bagged it and mulched with it, then spent another year pulling the bits out again.

Avoid weeds like this that spread from bits of the plant. Seeds shouldn't be a big problem because they get covered with subsequent layers of mulch. So long as you keep on mulching, they don't get a chance. I tend to figure that the seeds are floating (literally—on the breezes) around my area anyway, so I don't worry.

A neighbour stopped collecting other people's grass clippings because, she said, of dog droppings and chocolate wrappers. I haven't noticed a problem with the former and I'm still hoping to find a chocolate wrapper that isn't empty.

I have never, to my knowledge, imported a plant disease. Of course, it is possible to do so. You simply have to make up your own mind about whether or not the risk outweighs the benefits.

Avoid using material that has obviously come from diseased plants or from nearby diseased plants, whether it's in your own garden, somebody else's or discarded in the tip. For more information about garden hygiene, see *The Reverse Garbage Garden*, Chapter 11.

Won't it attract flies?
Not if dodgy materials are covered. Kitchen wastes and manures should go under bulky mulches like grass, hay, seaweed, straw or weeds. Then flies won't be able to reach them.

Fish or meat wastes can be buried in the garden, put into compost, completely covered by other mulches or cycled through the chook pen first.

Don't put on green lumps of wet grass clippings as these can produce maggots—thin layers don't. Dry materials like sawdust discourage flies and other insects.

Will mulch attract cats and dogs or rats?
Most mulches don't. In fact, cats like to use soft soil for toiletry purposes and mulch discourages them.

Dogs can be attracted by meat or by kitchen scraps. If fruit and vegetable scraps are well covered, most dogs leave them alone. If there are possible problems with dogs, don't put meat scraps anywhere near the surface. Burying is the best solution.

I have never had a problem with rats. Perhaps this is more likely in cities? If rats are a persistent problem, try composting kitchen wastes instead of mulching with them. A well-sealed compost bin might keep them out, or consider a tumbler-composter. These bins are suspended off the ground so that you can rotate them—I imagine rats would have trouble getting in.

I can't visualise other mulches creating a rodent or cat and dog problem.

Won't mulch attract insects, slugs and snails?

Maybe. Mulch certainly harbours, feeds and shelters insects. However, this includes predator insects so there is no reason to assume that only pests benefit. Mulch is also good for insect-eaters like frogs, lizards and toads.

I don't have serious pest problems although I've used loads of mulch for a decade in my garden. I think variety is the name of the game. Lots of insects eating lots of other insects plus attracting birds, frogs and lizards to continue the carnage. Aphids are a terrible nuisance in the garden, but did you know that there are alphids that eat other aphids? The only way to keep your predators happy and safe is to give them a good environment to live, breed and eat in.

Slugs and snails? Some mulches attract them. They love kitchen rubbish and hay—moist, deep mulch is their delight. They hate sawdust and wood ashes and anything with sharp edges or surfaces, so some mulch they like and some they hate.

I find the material they like very useful. It is easy to lift a slab of hay and harvest the slugs from underneath it. They cluster there. Many people set traps in their gardens to catch slugs and snails—half a grapefruit, perhaps. I don't bother because my mulch acts as traps. Logs and pieces of wood are excellent.

I had more trouble with slugs before I began using extensive mulch. My paths are made with dry, unpleasant sawdust (unpleasant to the slugs) and help prevent slugs from invading my garden areas. My use of mixed mulches means the slugs don't have it all their own way. Plus, I pick up solid bits of mulch and remove the slugs I find gathered there.

Can mulch attract termites?

This question came to me from a Western Australian gardener. I guess it can be a real problem in some places.

There are more than 350 species of termites in Australia, but only a few cause problems. I am grateful to the Agriculture Protection

Board of Western Australia for the following advice.

Pest species of termites are attracted to sound timber and are not much attracted to mulches like grass, hay, leaves, manure or seaweed. Problems can occur with cardboard, nut shells, paper, pine bark, sawdust and wood chips. Termites that are attracted by these materials can then move into timber and trees in your garden or perhaps into buildings.

The giant termite, *Mastotermes darwiniensis*, is particularly bad for fruit trees and vegetables and is found north of the tropics.

If you know that termites are a problem in your area, avoid wood-based mulches and nut shells. Take care generally. Keep an eye on timber in your garden as well as checking buildings.

Can mulch create a toxic build-up in my soil or plants?

There are three ways in which mulch can import toxins into your garden:

1. the material could have been sprayed with agricultural chemicals or herbicides;
2. manures, especially from feedlot or intensively kept animals, can include antibiotics and growth stimulants;
3. toxic amounts of chemicals or heavy metals may have been used in the manufacture of the original product.

1. Sprayed materials

The first category refers to materials like hay, plant wastes and straw, which are the products of conventional farming. Cut grass and weeds from the roadside and government land may also have been sprayed with herbicides.

If you use grass that could have been sprayed, ask about it. Ask the farmer, the shire, the workers, the neighbours. Follow it up until you are certain whether or not it has been sprayed. Although I don't believe that small amounts are going to do a lot of harm, it's best not to keep using material from sprayed areas. It's easy enough to look elsewhere.

Gardeners must make their own decisions about using produce which has been grown with the aid of artificial fertilisers, fungicides, herbicides and pesticides. I use it and feel sure it does no harm. The materials themselves add valuable organic matter and improve the soil so I'll take the bad with the good.

Since I use a wide diversity of mulches on my garden, only some of it has been exposed to these contaminants. If you believe that the bulk of your mulch is contaminated, it would be worthwhile looking around for alternative sources of material.

2. Manure from intensive farming

This is most likely to be chook manure, but can also refer to pig, sheep and even cattle manure if the animals have been farmed intensively. In feedlot farming, animals are kept in tiny pens and are fed high energy grain-based foods often mixed with chemicals to encourage rapid growth.

Most manure probably involves some exposure to antibiotics and perhaps to growth stimulants and the more intensively animals are kept, the higher that exposure will be. Reactions from gardeners are mixed on this issue. Some totally refuse to use farm manure.

Although I firmly believe organic food is better for me than non-organic, I nonetheless eat a lot of conventional food and don't worry about it. Standards are useful, but fanaticism creates problems and divisions. I presume that the good the food does in providing nourishment offsets the harm done by the chemicals used to grow it.

Similarly, I feel my soil and garden can cope with this indirect exposure to chemicals. The good the manure does outweighs possible and theoretical harm.

3. Materials that have had chemicals used in their production

This question often arises in reference to cardboard and paper, since heavy metals may be involved in their manufacture.

The evidence tends to suggest that these materials are safe to use. Levels of toxicity are very low and vary from one product to the next. Perhaps if you used cardboard and paper exclusively for the next few decades, you could experience a problem. Then again, perhaps not.

Cardboard and paper are rarely used on their own. They blow around easily if they are not covered by other mulches and have little in the way of nutrients to contribute. So—continue to use them with confidence as one element of your mulching programme.

Treated pine logs and sawdust are best avoided since they contain chemicals to inhibit decay and kill insects. Once again, a bit probably doesn't matter. There are a few treated pine logs in my garden, although I try to find hardwoods instead. With sawdust, ask the sawmill or tradespeople what wood it came from.

My response to this question, seen as a whole rather than in bits, is that a little of everything is all right. Using a variety of materials generally means the risk from any one of them is slight.

This answer refers to free mulches and to materials that are not specifically mulches like hay and sawdust. Commercial mulches are sold to gardeners and I assume they are safe.

Won't it make my soil acid?

Mulch generally *is* slightly acidic. However, the effect of adding organic matter to the soil is to buffer plants against imperfections in soil pH. Plants can put up with soil that is on the acidic or alkaline side better if there is a generous organic content.

Sprinkling wood ashes or lime sparingly over mulch corrects its tendency to be acidic. If you have an alkaline soil, and many Western Australian gardeners do, you might welcome this quality in mulch.

My heavy clay soil was acidic to start with. I have poured mulch onto it for the last twelve years, including leaves, pine needles and sawdust. Every now and then I sprinkle a few wood ashes around—not often because our wood-burning stove is very efficient. My soil has improved out of sight and my garden produces well.

There are mulches that are quite acidic, e.g. some leaves, such as oak and she-oak leaves and pine needles. Put these especially acid mulches around acid-loving or acid-tolerant plants, or make sure you spinkle them with wood ashes or lime to correct acidity.

Should I wash the salt off seaweed first?

Most gardeners who use seaweed don't wash it first. There is some suggestion that using seaweed exclusively and heavily for decades *might* possibly, perhaps, lead to a build-up of salt and the destruction of soil life. And perhaps not. Anyway, who uses seaweed that way?

I have read a suggestion that seawater be used as a mineral supplement. The idea was to get a jar or bucket full, dilute it a lot, then spray it on the garden. Old time gardeners used to use salt as a fertiliser on asparagus, believing that this plant of seaside marshes needed it. So, sea salt can't be all bad.

The small amounts of salt that cling to seaweed aren't going to do any harm. It is perfectly safe to use unwashed seaweed as part of your mulching programme.

How can I tell if my plants are healthy?

I don't have any inside knowledge about this. If plants are a good green colour (I mean those bits that are meant to be green—green carrots might worry me), are growing strongly and generally look good, I assume they're healthy.

Discoloured or mottled leaves can be a sign of disease or of mineral deficiencies. However, I don't waste my time worrying about a few mottly leaves. If the condition is spreading between plants or the whole of one plant succumbs, then I do worry. My normal response is to remove such plants and burn them.

Yellowing and pale leaves can be a sign of nitrogen deficiency. Plants need nitrogen constantly to make good growth. If you suspect this condition, sprinkle blood and bone around the affected plant or garden bed, or water with manure tea or well-diluted urine. Manure tea is made by soaking manure in a bucket of water for a week or two, then diluting the mixture until it is the colour of weak tea.

Wilting on a hot day indicates water stress. When a plant wilts, it hangs down limply. Usually plants pick up again in the evening. However, wilting is a sign that your plants are suffering stress from lack of water. If you have water, give it to them. It is best to avoid this situation by preventative watering in hot weather. Plants can still suffer water stress in very dry weather, even if they are well mulched.

If none of this is happening, you can safely assume that your mulch is keeping your plants happy and healthy. For further information about plant deficiencies, see *The Reverse Garbage Garden,* Chapter 4.

Should I cover mulch with plastic?

A non-gardener asked me this. I'm not sure if he was confusing mulch with compost, or thinking of plastic put down as mulch.

When plastics are used as mulch they are put on the ground to conserve warmth and to stop weed growth. Because plastic has no nutrients, you can put high value mulches underneath it to provide food for your plants and organic matter for your soil.

However, plastic is only used by some gardeners and often in very specific circumstances. Professional strawberry growers and some market gardeners put plastic along their rows. They combine it with the use of artificial fertilisers. Plastic is also used where soil fertility is less important than weed control. In neither of these cases are other mulches put under the plastic.

Gardeners who use plenty of organic materials don't cover them with plastic.

Mulch is often put *over* plastic in gardens where weed control and landscaping are the major concerns. Long-lasting materials like wood chips are used. This is not suitable for fruit or vegetables since it doesn't provide nutrients.

Compost heaps are often covered with plastic to keep them from drying out or getting excessively wet from rain and to help them warm up. But compost is not the same as mulch.

10

Gardeners Tell All

When gardeners are asked to bare their souls, what do we find? A dark, rotting mess filled with worms and toads!

So far the voice in this book has been mine. It's time to give someone else a go. I asked people, most of whom I'd never met but who I knew were organic gardeners, to tell me about mulch. I did ask for negative experiences as well as positive ones, but not a lot have come through.

I tried to contact gardeners from around Australia, but have to admit that my sample mostly comes from the eastern states. If any of you Territorians, South Australians and Western Australians would like to write to me, please do.

I would like to very sincerely thank all the people who agreed to share their experiences in this way. I have given first names only to protect their privacy.

I have, for no reason, decided to work downwards from north to south, Queensland to Tasmania. Then I jump westwards. I have included postcodes so you can tell in what part of the state the gardeners live and therefore what the climate's like.

I hope that reading these letters will be like having a whole bunch of fanatical mulchers talking directly to you. I hope that hearing their voices is going to inspire you so much that you can hardly wait to get your hands on some grass clipping or leaves or seaweed or whatever.

Go get 'em!

Sandi from Ravenshoe, Queensland 4872

We moved from a dry tropics area to a Tablelands area so I'll tell you about both.

On Cape York Peninsula, which is a dry tropical region with hot, humid, wet summers and dry, warm winters, we have a six monthly dry period, with the ground very dry by October. The weather is getting hot by then and the natives tend to shed their leaves.

Gardening in this climate is the reverse to down south, with summer things planted in April–May towards the end of the Wet then irrigated as it dries out. The Wet (approximately December–April) is our off season as far as conventional vegies go though there are New Guinea and Asian types that grow at this time.

Mulching is useful to prevent erosion when it is raining heavily. We did have a problem, though, in that it provided an ideal breeding haven and shelter for fire ants—tiny red ants that bite like hot coals. They are very destructive to seeds such as beans and peanuts. Where we turned garden areas over to poultry the fire ants diminished drastically.

The mulches we used were locally cut blady grass which decayed rapidly and didn't provide a lot of nutrition, kitchen waste, newspaper and cardboard, old clothes, potato sacks, rainforest leaves, scrub turkey nests (excellent mulch) and lots of dolomite.

Partly decomposed rainforest leaves were placed around tropical fruit trees about 30 centimetres thick. This was never really thick enough but everyone would get tired of hauling bags of leaves all day.

Now I'm on the Tablelands, 1000 metres above the coast near Cairns in the wet tropics region. It's cooler and drier here than on the coast but has similar weather—hot, wet summer and dry winter (but much cooler; we even get frost here occasionally).

Because it's cooler we grow temperate climate things like stone fruits, grapes, cabbage, lettuce all year, even apples. Strangely enough, we can also grow bananas, mangoes and pawpaws but not pineapples.

The school mulches with sawdust, it being a timber area. Further north in the Atherton area, peanut shells are used extensively. Councils use them for parks and street gardens.

I am on a commercial, organic fruit and vegetable farm and we use goat manure, chook manure, spoiled hay, donkey manure and leaves and kitchen scraps. Stuff that stays awhile like rhubarb and comfrey is mulched, other areas are just fed the manures. The fruit orchard is mulched from time to time with manures and rock dust.

We just don't get time or enough hay to completely mulch the garden all the time. Spoiled hay is also put in the chook house, then shovelled onto the garden when it needs changing.

Keith and Maree from Kirwan, Queensland 4817

Our no dig garden here in North Queensland has had its ups and downs. The biggest down is that with both of us working any type

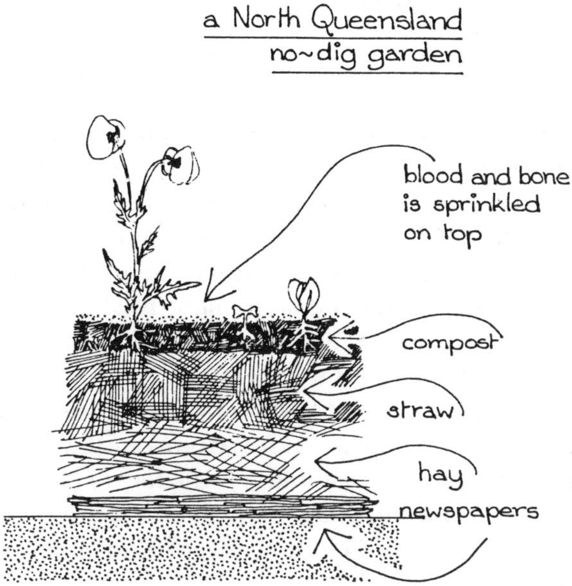

of garden requires time to be put into it. And alas that is the thing we have least—spare time.

The second thing is up here we have the most amazing array of weeds imaginable and in particular nutgrass is the thing we constantly battle—by hand—no pesticides.

Our no dig garden was constructed from:

1. a *thick* layer of newspaper;
2. a thick layer of lucerne hay;
3. a thick layer of straw;
4. a thick layer of our own compost;
5. blood and bone sprinkled over the top and watered in.

The type of vegetables planted here during the year has to be carefully planned as the heat would kill a lot of seedlings before they had a chance.

Some people rig shadecloth over their vegetable gardens or part of the plot at least.

Lynn from Brisbane, Queensland 4017

My problem is nutgrass. I started to mulch with thick layers of newspaper and cardboard and old cotton towels and cloth—anything I could get my hands on. I covered this with a thick layer of straw. Where I grow native and fruit trees, the nutgrass is the worst, so I have given up growing vegetables there.

I have dug a vegetable garden right in the middle of the backyard.

So far I haven't had any problems with the nutgrass there. The newspaper and straw seems to make a good barrier. I have noticed that nutgrass comes through cardboard more easily than through newspaper.

I found the nutgrass started to sprout again where the newspaper had decayed. I have started raking the straw back and putting down a layer of newspaper, twenty sheets thick, and then putting the straw back. I think I'm starting to get on top of it now.

We have over thirty trees on an ordinary suburban block. It's easier to mulch around them than to mow. I use newspaper and straw—wheat or barley and occasionally lucerne hay which is more expensive. I buy them from produce stores.

In the past I have also used hessian sacks, sawdust and wood shavings, which I was able to get free from a mill. On the vegetable garden I use straw and have found lucerne hay to be excellent. So much so I have decided to plant a few square metres of lucerne in a corner of the garden. I will cut it several times a year with a bruchcutter.

On the rest of the garden I use shredded prunings from native trees and shrubs and leaves. Small leaves are put on unshredded. In the past I have used commercial pine bark but some of it still hasn't broken down after ten years. I prefer the look of the native mulch and after about twelve to eighteen months most of it has disappeared. There are always lots of worms underneath it.

I have two narrow beds out the front where I plant annuals which soon cover the ground. I don't mulch these.

The mulch helps keep the weeds under control and the soil does not dry out quickly and is not washed about so much during heavy downpours. An added bonus is the number of tree seedlings that appear in the mulch. I pot them and plant them on a few acres we have in the Conandale Ranges.

Most of the fruit and vegetable gardens that were here when we came nineteen years ago have gone. The population has aged, older folks have died, and many houses are now bought as investments.

The Brisbane City Council has a recycling programme where garden waste is shredded. They use mulch on parks and gardens and around trees planted on footpaths. When my boys were at primary school the garden beds were all mulched with wood chip and pine bark.

Jeff from Argenton, New South Wales 2284

Jeff grows and sells herbs under the name Hozy's Herbs.

The local councils, state authorities and other bodies mulch extensively in this area. Trees that are pushed down for road works, trimmings from powerlines and general prunings from public places are run through a portable wood chipper. This is stored and used as required.

On new freeways wood chips are laid down and natives planted on top of or into this. It is astounding just how quickly the natives grow. One drawback as we found out last month (January 1994 bushfires) is that the mulch does give bushfires a ready supply of fuel if it has not rotted down enough.

The most obvious reason for mulching here is the fact that mulch reduces the amount of water needed. In the Hunter Water Corporation area we pay for every drop of water we use so this is very important—save water, save money.

The other main reason is to keep down weed growth—why pay for water to grow useless weeds? I find newspaper very useful. I lay down a bed of paper, which looks untidy but I put lawn clippings or anything I've got over the top.

Mulch also warms the soil to speed seed germination and it increases the worm population thus improving the soil. Weeds or crops I pull out are left lying, roots upwards, to dry out and rot down to add humus to the soil.

Possibly the best mulch is cottonseed hull (or husk). It is dense enough to stop weed growth, open enough to allow water to penetrate and it is neat and clean. It is also fourteen per cent protein and can be turned into the soil when the crop has finished.

I discovered a supply of old hessian bags that were being dumped each week at the tip. I rescued 500 to 600 before I ran out of space. I use these as mulch too.

Hugh and Lilian from Mount Colah, New South Wales 2079

This married couple wrote separate letters and, in fact, maintain separate gardens. Hugh, who has long been associated with the Henry Doubleday Research Association of Australia, Inc., has some very decided views.

Lilian

Getting enough mulch was a problem until about eighteen months ago when a young chap who cuts lawns for a living was directed to us and now he brings the grass clippings here. He cannot figure out how we use so much of it. Sometimes it has lots of leaves chopped up through it and as it dries off it makes marvellous mulch. Sometimes I leave it until it is almost soil and full of worms. I find this a real godsend.

I have used heavy newspaper mulch topped with grass clippings in various parts of the garden over many years and intended to do it everywhere. In our light sandy soil it is difficult to retain moisture. Without mulch we could give up gardening here. I put heavy mulch around vegetables, sometimes with well rotted animal manure mixed in.

I have also brought home from work shredded computer paper and put it around plants. This works well and seems to reflect a lot of heat.

A friend has wonderful success with her vegetable garden by using clumps of wool from discarded fleeces from the sheep on her property. This is wonderful insulation for the soil. The main thing is to mulch with whatever you can scrounge and find out what works best for you.

Hugh

My plants are planted in beds of properly made compost. I have three compost heaps—one is being filled up, one is 'cooking' and one is being used. Everything that is living or has lived goes in—household scraps, waste water, coffee grounds and tea slops, but the main ingredient is weeds. The goodness and strength for your compost comes from weeds, especially deep rooted ones.

I don't put any kind of green plants on the garden as mulch as this is the worst thing you could do. I make very sure anything I use as mulch is very dead and very dry.

I use almost anything that is at hand as mulch, but what I like best is long, dead, dry grass such as Kangaroo, Wallaby and Whiskey grass. We gather it from the roadside when we drive out of town.

Organic gardening without mulching is unthinkable; you only need a fraction of the water you would use without it, it warms the soil in winter and cools it in summer. When the bushfires came so close to us a few weeks ago (January 1994 bushfires) I didn't lose one plant.

Using long grass and lucerne or any other dry spoilt hay as mulch will cause seeds to grow up in the garden, but no matter—they can be pulled up and put in the compost heap.

We live in the Hornsby Shire. They put all their tree prunings and so on through a huge mulcher and use the green mulch on flower beds in parks. This is one of the worst things you can do—it should be allowed to rot down or compost first.

Neil from Briagolong, Victoria 3860

I have used an enormous amount of old hay obtained for free, people bring me their lawn clippings and I have raked up heaps of leaves over the years. I don't use cardboard because I have been told it has a high aluminium content and I use a minimum of paper.

I collect sawdust and wood shavings from a local wood turner and use these on the paths as well as for a light mulch. I also have a mechanical shredder and have put through a lot of wattle.

At its peak my garden was about 700 square metres and very productive.

The two things that cause weed seed to germinate are light and disturbance. Once seed has started to germinate it will grow through a reasonably thick mulch. However, if seed has actually germinated and is at the 'two leaf' stage, a thin mulch will suffocate it.

You can disturb the soil, wait a few weeks till the weed seed has just germinated, throw desirable vegetable seed about on the bed and apply a medium thickness mulch. The mulch suppresses the weeds that were already growing but the vegetable seed germinates and comes up through the mulch. This works well with large-seeded vegetables.

I also make quite a lot of compost because locals bring me far more grass clippings than I can use. Sometimes I have to ask them to stop giving it to me!

Di from Woodend, Victoria 3442

We have a clay soil here, not that I think it's a bad thing as clay has a lot of nutrients. We didn't have a tree on the block when we bought it, so it was an awesome thought starting a garden from

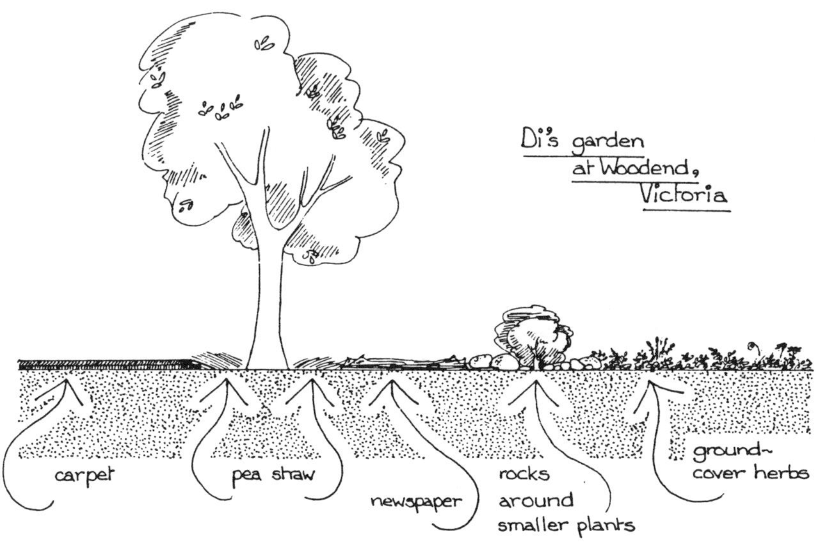

scratch. Luckily I had read Bill Mollison's *Permaculture* and Esther Dean's *Growing without Digging*.

I use newspaper or carpet on larger areas around trees but not right up to the trees. I don't think it lets the water through well. Then I use pea straw to the base of the tree. It rots well. We also use bark off the firewood we collect—it is more decorative so it's used in flowerbeds and the like. Smaller plants can be propped up with rocks of different sizes. They give shade and retain moisture around the roots.

I use a lot of herbs as ground covers. They're hardy and grow quickly. Strawberry plants are terrific for this. They give fruit and propagate easily.

I don't buy mulch but a nearby town sells a beaut mulch made from recycled garden throw-outs from the tip; it's a lovely dark mix.

A lot of people I know use the paper and straw method, maybe raking leaves on top and adding animal manure. I buy large bags of horse manure on the side of the road nearby. I don't have any trouble with weeds (maybe one or two). I have chickens and make a chook poo and water mix that goes on most of my plants except the natives.

Ruth from Bell Park, Victoria 3215

I live close to both bay and coastal beaches. I pick up seagrass from the bay beaches nearby. It is a very fine seaweed, different from the thick coarse kelp and others found on coastal beaches.

At first I carefully hosed it down to wash out the salt but now I simply put it straight on the garden and I haven't had any

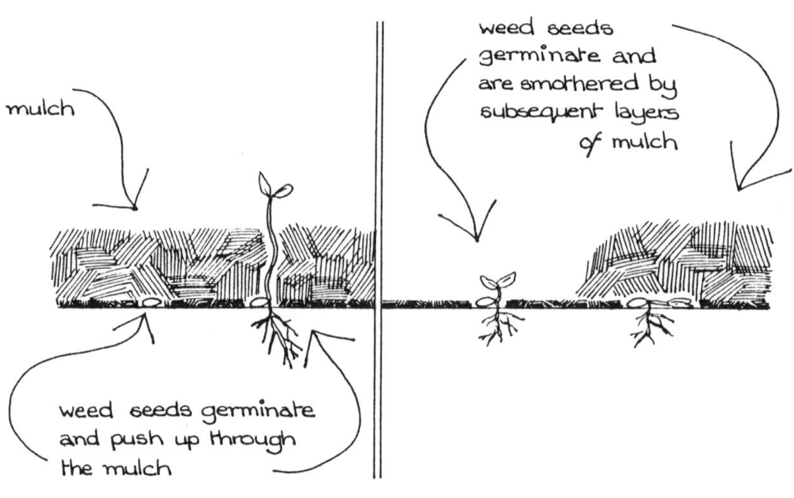

problems. For the first few times that I water or it rains, it smells a bit like a beach but the smell is pleasant and soon goes away.

The seagrass contains some sand, shell grit and shells which not only look nice but add minerals to the soil.

Seagrass goes a long way because a layer of about 5 centimetres is all that is needed. As it is slow to break down it only needs to be topped up every second year or so. It forms a mat over time and it is rare for a weed to find its way through. It is pleasing to the eye and all it costs is the effort to collect and transport it.

I recommend seagrass to anyone who has access to it.

Sue from Kongwak, Victoria 3951

Sue is the secretary of an organic growers' group

- Seaweed: easy to get, although we have to get a licence first, easy to spread, organic and effective in blocking weeds. It is heavy to handle when wet.
- Straw: looks good, suppresses weeds if heavy enough, easy to spread. It is not organic, may have seeds and costs money.
- Hay: organic if it's our own, suppresses weeds, easy to collect and spread. It contains weed seeds.
- Pine needles: organic, looks good, suppresses weeds. It is acidic and we don't have a source nearby.
- Sawdust: suppresses weeds, looks good, easy to collect and spread. I don't normally use it on its own.
- Compost: we spread it thinly on the soil and put other mulches on top. We don't have much of it.
- Cow manure: organic because it's produced on our place, free. It isn't easy to collect, can be messy and smelly and it spreads weeds. We also use duck manure mixed with hay from their cage.
- Wood ashes: I tried this around seedlings because of slugs and snails but I found that the only answer was to go out for five nights or so and kill them.
- Newspaper and cardboard: suppresses weeds, freely available. It's not organic and will blow away unless covered with something. We use it for paths, around the edges of garden beds and right through the native garden area with mulched tree prunings we get from the SEC (State Electricity Commission).
- Weeds: we used your method (see *The Reverse Garbage Garden*, page 100) of growing potatoes in pulled-up weeds and it worked fantastically and was so simple. We try not to use the deep rooting grasses that send out shoots all over the place but put other weeds in between rows of vegetables.

Sometimes I use mulch in certain combinations for no logical reason. I think it can be an intuitive, creative experience!!

Robyn from King Island, Tasmania 7256

It is seaweed gathering time here right now, so this is a good time to write about mulch.

Last year the mounds of seaweed were at least a metre high and they remained on the beaches for weeks. Seaweed is not nearly as abundant this year. This is on the eastern side of the island and is different from the kelp that is harvested and dried for processing in Scotland.

This is fluffy, tendril, or leaf-like stuff; an amalgam of the fairy clothes lines, sea necklaces, sea-lettuce, poppers and whatever else comes up after a good easterly blow. It usually only comes once a year. Then we all troop out with our old super(phosphate) bags, collected from the local tip, and bag it up.

The beach is right in front of my house, but there is a steep climb up to my garden, so I tend to drive a bit further down the road to get it.

Nearly everyone with a good garden uses it for vegies and flowers. Nobody washes it before putting it on and they all swear by it. One side of the island has light, black sand (my side), the west has yellow calcium sand and the rest has heavy clay but seaweed helps the whole lot. I am not much of a gardener but things grow.

Another mulch available to island gardeners is wool dags which can often be collected from the tip. Farmers let you dig the droppings from under the woolshed gratings too and cow manure is available from dairy farmers.

Lately we've had mulched up trimmings available from where the roadsides are kept clear under the powerlines. It consists of chunks of native bush—ti-tree, wattle, banksia, gum, etc., and it includes leaves and branches. It is more a covering mulch than a soil improver but it looks good under thick plantings of native shrubs.

Most people here depend on tanks for water and, as it is dry in summer, mulching is important to King Islanders. I bring home the groceries in cardboard boxes, save my newspapers, and magazines and old carpets come from the tip. Old seagrass matting is good too. I even use old vinyl flooring under my vegie boxes and on pathways, but it can be slippery in the rain.

Bracken fern is good if it is chopped in a mulcher or slashed—it makes a cushiony mulch that breaks down over a year and doesn't usually bring weeds.

Hay is normally available, but I find it grows too much of itself.

I often can't believe just how easy and good mulching is!

Bob from Cygnet, Tasmania 7112

Bob sells dwarf fruit trees via mail order. From previous correspondence with him, I know he mulches extensively around his trees and plants under and among them.

In our thirty years of horticulture, mulching has always been the highest priority, especially here where we have very heavy clay soil that was flogged for eighty years before we got here.

I have a source of mulch that I like to keep secret as in the past I told many people and suddenly there wasn't enough left for me!

In Tasmania about the most popular mulch is pea straw which comes from the north-west coast where they grow peas for processing. It's great light mulch with a bonus of a crop of peas—you buy it in bales. Seaweed or seagrass is also popular amongst the cognoscenti and there are several east-facing beaches where you can usually gather some.

Private schools and church groups sell bags of sheep manure. Poultry farms and horse stables also sell manure but some people fear these have too many antibiotics, etc., in them.

However, I'd like to say that though we have a mulch ethic here, it's not usually carried through. People complain about their soil but never do much about it.

Nonie from Perth, Western Australia 6050

We live in Perth, an area of very low summer rainfall, so water conservation is a very important but often not considered part of gardening and farming.

The kind of mulch we have used has always been whole tree prunings put through a muncher. These are bought cheaply by the truckload from people who prune trees and consist of a variety of tree types, unlikely to have many pesticide residues or weed seeds. It is a healthy combination of broken down leaf matter and longer-lasting woody chips.

In summer we broadcast straw over the mulched areas—the light colour and airy spaces in straw act as reflection and insulation to reduce evaporation and heating by the intense Perth summer sun. The straw breaks down over winter. Loose straw also provides a niche for useful insects and helps reduce wind damage to small plants.

Much of our garden has been sheet mulched. A layer of wet newspaper is placed over whatever we want to get rid of (couch grass, kikuyu, buffalo grass, bare sand, weeds and weed seeds) and then on top of that goes mulch such as munched prunings, manure, straw, leaves or lawn clippings. We have had varying success with this method. I can make the following observations:

- Sheet mulching with newspaper is okay in areas of consistent rainfall and watering. Otherwise the paper doesn't break down and the soil can become compacted.
- Biodegradable carpet underlay can replace the paper as it breaks down faster—but this is no good against couch grass.
- Manure under and above newspaper helps it to decay. Putting on active compost helps; so do worms and slaters.
- Larger plants should be put into the soil. Make a slit in the newspaper with a hatchet.
- Kill couch grass or kikuyu first by covering it with corrugated iron for several months or use thick newspaper and lots of manure.
- Zolla (*Azolla filiculoides*) grows profusely in a garden pond. It is nitrogen fixing and makes a superb mulch. Scoop it out of the pond and spread it.
- Mulching with whole branches creates niches for young plants and wind shelter and approximates a natural system.
- Kelp, collected from the beach and rinsed, is full of trace elements. It lasts for ages but breaks down quickly in a compost heap.

Finally—there is no one method that is right for everybody. We must find what's appropriate and be prepared to learn from everything we do.

11

Who Needs Mulch?

This is a deceptively simple question and the answer peels off in layers.

At the simplest level, I can say that *all* gardeners will benefit from mulch and therefore *should* be using it. The use of 'should' is really a rhetorical device for emphasis. I don't believe any of us ought to preach or berate or attempt to instil guilt.

Action, not proselytising, please.

THE NATURE OF ACTION

On another level (peeling off a few layers), political action is called for as well as hands-on, grubbing-in-manure action. There's all sorts of shit that needs to be stirred, after all!

Our local shire uses a hired chipper to process branches and trees it has to cut down, but only because of the dedication of a particular councillor who pushed hard for this project. Without her, gardens in parks and through the streets might not now be mulched and timber removed from the roadsides would just be rubbish.

Hers was political action, motivated by her views about renewable resources and conservation. Between her job and her unpaid work as a councillor, the poor woman probably never even has time to get out into her garden let alone to mulch it.

Without political action, change is slow and may never happen. We have to have actors (those who act, who do) in the garden and actors in the fields of decision-making tall poppies. One hugs the ground, one flies high—both are needed.

Political action is necessary at all levels, from local council and shire to federal government. Political action doesn't just mean becoming a councillor or a member of parliament, it can also refer to writing letters to the newspapers, ringing up (relevant!) radio talkback shows, publishing books, writing songs, etc.

Propaganda and political action get mixed up, unfortunately. In forcefully presenting the case for mulching (and similar sorts of action), it would be nice if people could retain their sense of balance. Your first responsibility is not to (smooth) talk or bludgeon people into mulching, but to maintain ethical integrity. Nonetheless, I'm a pragmatist and recognise that propaganda will be used.

Different people are likely to work politically and practically. It suits some people's aims and ambitions to get out there and mix it with the best. Others, with similar motives, prefer to 'potter' in their gardens, creating change by example.

Both are valid ways of acting.

GOVERNMENT AND OTHER ORGANISATIONS

Mulching is valuable for organisations because it simplifies their gardening, reducing the need for watering and weeding. The main labour is in the establishment of a bed, rather than the maintenance.

Plastic can be used as an effective weed barrier. Bark or wood chips on top prevent wind-blown seeds from germinating for a year or two. Combined with a careful selection of plants, mulching eliminates the need to water.

Closely spaced plants soon grow together, completely covering the ground between them. This can mean no more mulching *or* weeds. The plants shade the ground themselves (robbing potential weeds of sunlight and conserving soil moisture) and provide a scattering of their own leafy mulch for soil improvement.

Council and office buildings, factories, car parks—these are the 'impersonal' sorts of gardens that benefit from the way that mulch dramatically reduces garden maintenance. In these situations, maintenance equals money: save maintenance and you save money. Mulching is cheap and more than pays for itself in saved labour costs.

Erosion control is an important factor along highways, freeways and roads, and I have seen mulch use increased in these areas in recent years. Banks are often steep. Because of the landscape creation that major roads require, the soil is not exactly ideal for plant growth. The mulch stops the soil sliding down in rain, allows plants to get going in poor conditions and eventually helps improve the soil.

Councils also use mulch for erosion control in parks, beside creeks and rivers and on any sort of bank. An employee of Richmond City Council (Melbourne) told me they find euca mulch (see page 71) very good as it mats together to protect soil on steep slopes.

We should all encourage organisations, especially but not exclusively government organisations, to mulch gardens because this gives the right message (as far as we are concerned!). When ordinary citizens see responsible bodies mulching, it is an encouragement to them to follow suit. The message is: mulch is good.

The organisations themselves benefit from weed control, water conservation, erosion control, and a decreased need for maintenance and labour input. A well-established, mulched garden with suitable plants needs very little maintenance.

Organisations with a social need for gardening

Some organisations encourage human input into their gardens. Hospitals and schools sometimes do this.

Mulch is useful here as well, because a garden that has been established by either professionals or volunteers is then easily maintained, fully or in part, by volunteers.

Kids can enjoy gardening and are often asked by primary schools to help as part of the extra-curricular activities. However, they lose heart and interest readily so they need good results in a hurry. Mulch can keep kids' gardens neat and productive. They easily learn how to put plants into the soil under the mulch. I can assure you, kids don't enjoy weeding!

If a school wants to encourage some food growing, it is easy to transfer mulching skills and knowledge from ornamentals to vegetables. Useful messages are being passed on about conservation of resources and reuse of wastes, especially if the school shows kids how to put the school's own cut grass on the garden.

Patients in hospitals and elderly people in hostels (which may be attached to the hospital, as in my town, or be separate) can gain a lot from gardening. Being outside is healthy (subject to doctor's orders) and they can feel useful and valued and may also quite simply enjoy the activity. One of the difficult things for the elderly or temporarily incapacitated is living with idleness. We *need* to do things.

For these people, gardening should not be too strenuous or difficult. Mulch reduces the need for heavy labour in growing either ornamentals or vegetables so it is ideal. I know of patients who have established vegetable gardens at hospitals, sometimes drawing employees into the fun, so this is not just a fairy tale.

Residents should be encouraged, as much as they can, to participate in all the gardening. If someone loves roses, why shouldn't they be helped to start a rose garden? All kinds of gardens will still be there decades later if they are mulched. Even a vegetable garden, abandoned because its creator has died or gone home, can be turned into a mulched ornamental bed. Its soil will have been improved by the work done on it.

A friend, whose job is organising accommodation for homeless or troubled young people, is starting a vegetable garden in the backyard of a special accommodation house. Even though the kids will come and go, the garden will remain, providing some fresh food and a sense of achievement. It was *The Reverse Garbage Garden* that inspired my friend!

Organisations can use mulch to make the work easier, more fun and more readily productive.

INDIVIDUALS

Naturally, I think everyone who has any sort of a garden, including one in a pot in a second storey apartment without even a balcony, would benefit from mulching. However, there are certain people who need mulch more than others.

The elderly

Older people may enjoy plants and gardens and still want to have their own, yet be held back by physical disabilities. They may suffer arthritis, rheumatism, impaired vision, decreased mobility and will almost certainly have a fair few minor ailments.

Using raised beds and mulch together can create a low-maintenance garden you don't have to stoop to. There are volunteer organisations in major cities that set up such gardens. Elderly people can then carry out most of the maintenance themselves, easily sowing, planting and transplanting. Ornamentals, vegetables and small fruiting plants can be intensively grown in this way, according to taste.

No bending or digging is needed. Mulch can be bought or supplied by volunteers, then stored and used by the gardeners as they want it. Blood and bone plus wood ashes or lime combined with mulch will keep the soil in good condition.

Wide paths around raised beds even allow access to those confined to wheelchairs.

Ornamental gardens can be set up and maintained around flats and houses that elderly people live in. A friend of mine is eighty-seven years old and still has an active interest in her garden. Her small front garden is mulched from fence to house with plastic under scoria. Tidy little bushes dot the yard. It would look better (but I prefer controlled or uncontrolled jungles) if it were more fully planted, but it is neat and functional. It is also easily maintained by a gardener. My friend doesn't need to water and the gardener doesn't have to come often. Since he charges by the hour, the mulch saves her money.

Mulch can help the elderly by saving them money if they employ a casual gardener and by allowing them to maintain productive intensive gardens they can manage with just a little help.

The disabled

Physically or mentally disabled people may wish to garden. As with the elderly, raised beds can be set up to allow wheelchair access and to eliminate the problem of weeds invading from pathways.

Everyone can be taught to mulch. A little bit more or less mulch won't wreck the garden—there is, in fact, no 'right' amount of mulch. Erratic application only means that some weeds will grow and they are easily smothered by more mulch. You can't do any real harm with mulch.

With or without help, many intellectually handicapped people can maintain mulched gardens.

By eliminating the hardest work—digging and weeding—mulch enables many physically handicapped gardeners to cope. As with the elderly, they can spread their own mulch even if they can't collect it. Mulch also helps blind gardeners handle the complexities of unsighted gardening by eliminating weeds.

The blind or intellectually handicapped gardener can appreciate the textures of different mulches—an undernoticed phenomenon.

Watering is far less critical if a garden is mulched because the plants need less. If a gardener doesn't understand when to water or is physically incapable of watering at the time it's most needed, mulched plants are unlikely to complain.

Children

A lot of children have a sporadic interest in gardening. They are normally capable of spurts of immense energy but rarely of sustained effort in the face of slow returns.

Kids need gardens that are easy to set up, easy to maintain, and quick to give a significant return. They also deserve to know about conservation and recycling—not by lectures but through action and practice.

Your children are more than capable of looking for mulch, learning to recognise it, helping to collect it and spreading it on their own beds or around their own shrubs and trees.

Weeding is not fun. Kids soon tire of a garden that has to be constantly weeded. Standing with a hose might be interesting the first time or two, but scarcely after that. Save them the trouble and show them how to use fine mulches around seeds and seedlings and coarser mulches around larger plants.

Children readily learn the very simple basics of mulching: don't put moist or high-nitrogen mulches near plant stems; don't let mulch overwhelm seedlings; move mulch aside to sow seeds; use constant thin layers of mulch. Let them help decide what mulches to use where. Get them to make decisions about where and how to collect mulch.

Mulching ensures that a garden receives a mixture of nutrients. Adults can supervise the use of blood and bone and lime or wood ashes. In the case of very small children, the adults should spread these themselves. Older children don't need or want you looking over their shoulders.

Make sure you let the garden be theirs. You're just the adviser.

Because mulch can cover weeds completely, a garden can be

started instantly. Now! Not next week or next spring when the soil is soft. This prevents kids from losing their initial impetus.

Mulch quickly kills the weeds in a new garden and gives seedlings and plants their best chance to get going. This provides the fast results the kids are looking for.

Mulch can help make a kids' garden the happy success it should be. It kills weeds; gets the garden going quickly; provides nutrients; reduces watering; and it makes gardening easy. It also provides a lesson in practical reuse of waste materials.

Busy gardeners

And who isn't?

You, me, the guy next door and the woman in the blue dress—we're all busy. Whether you're at home all the time, are employed part-time or full-time, the odds are *you're* a busy gardener.

Mulching is ideal for *you*.

An ornamental garden can be mulched and planted out in a matter of days if you have material available in bulk, whether it is free or commercial mulch. Properly made, such a garden can be largely ignored for the next year. By then, your plants will have grown and spread. They can be designed and planted to cover the bed completely and mulch themselves from then on.

If good and improving soil is necessary, as in a fruit and vegetable garden, you can't be as casual. Mulches have to include some high in nitrogen and have to be suitably mixed. A food garden, in any case, requires sowing, planting, transplanting and harvesting, so it cannot be neglected. Mulching at least minimises the work required. Mulch may be the only thing that makes such a garden possible for you.

A well-mulched garden can feed your family with only hours of work a week. Of course, you *must* keep on seeking, collecting and applying your various mulches.

If commitments leave you too little time to dig, weed, water and otherwise fuss with your garden, mulch is the way to go.

WHY *YOU* SHOULD USE MULCH

You can probably answer this question better than I can! If you are not at least interested in mulching you're unlikely to have read this far.

By mulching you'll be doing your soil a favour—providing organic matter and nutrients for plant growth; increasing soil life;

protecting the soil from sun and rain—and you'll be doing *yourself* a favour. For very little work, you can have a terrific garden requiring minimal maintenance.

If you pay for water, and the odds are that most of us will soon have to if we don't now, you will save yourself money. Water = money. Save water = save money.

You can save yourself virtually all the hard work involved in gardening—digging and weeding. You exchange these back-breaking jobs for the relative comfort of bringing in mulch and throwing it on or patting it delicately around plants.

Many problem weeds can be successfully conquered by mulch. Deep rooters like dandelion and dock simply don't stand a chance. Mulch laughs at them. Couch, nutgrass, sorrel and twitch are tougher (see page 56). However, they're a bind no matter what you do. Very heavy mulching with good covering material like cardboard, carpet, cloth, corrugated iron, paper, plastic and pieces of wood are at least a big help. Maintain this mulch, combine it with constant vigilance against the foe and you should end up, if not victorious, then unvanquished.

Even if you live a metre from your neighbours in the inner city, you can still use heavy duty mulches to get rid of stubborn weeds. Simply make sure you cover unsightly material with neater mulch. Nobody can tell there are cardboard boxes, torn tea-towels and bits of wood underneath that neat layer of straw. Mulch can be used to create a weed barrier right around a small garden if necessary.

Mulch means that all the nutrients and water that are available go to your plants, not to weeds. You save work, time and money and your plants grow well.

Mulch helps you in a multitude of ways but, as well as this, there are other reasons for *you* to mulch.

When you start or expand your mulching, you are accepting personal responsibility for the fertility of your soil and for the health of your garden. Responsibilities mature and mellow us. Just ask any parent (wait until they're not screaming at the kids).

If you choose to use free mulches, and I hope that everyone will use *some*, you are taking a positive step towards solving our planet's waste problem. Instead of just leaving it to the journalists and politicians, or of dropping into incurable despair over the size of it all, you are actually *doing something about it.*

This is not only good for the planet but good for you, too.

By mulching, you set a good example to others. True, that good example is seen by few people, but that's the nature of changing by doing. You do it, others notice, talk about it, consider it. Some of those others copy you (more or less, in their own ways) and start their own little spheres of influence. These ripple outwards.

So go to it—it's a dog-eat-dog world, but instead of eating the dog, let's throw it on the garden.

Books

Clayton, S. (1993) *The Reverse Garbage Garden.* Hyland House, Melbourne

Dean, E. (1977) *Esther Dean's Gardening Book.* Harper and Row, Sydney

Hills, L. D. (1964) *Pest Control Without Poisons.* Henry Doubleday Research Association, UK

Mollison, B. and Holmgren, D. (1978) *Permaculture One.* Transworld Press, Melbourne

Morrow, R. (1993) *Earth User's Guide to Permaculture.* Kangaroo Press, Sydney

Murphy, D. (1993) *Earthworms in Australia.* Hyland House, Melbourne

Roads, M. J. (1989) *The Natural Magic of Mulch.* Greenhouse Publications, Melbourne

Yepsen, B. Jr. (ed.) (1978) *The Encyclopedia of Natural Insect and Disease Control.* Rodale Press, Emmaus, Pennsylvania

Mail Order Suppliers

MULCH

Earthworm Castings, 14 Parkland Crescent, East Brighton, Vic. 3187. Enquiries: National toll free (016) 37 6700.
Fauser Worms, Reply Paid 46, PO Box 400, Eastwood, SA 5063. Enquiries: (008) 88 2811
Lahuge Organic Services, 19 Bourke St, Dubbo, NSW 2830. Enquiries: Freecall (NSW only) (008) 81 7904.
Multicube ('Grow 'n' Mulch Cubes'), 'Warrawee', Gulf Rd, Yarra Glen, Vic. 3775. Enquiries: (03) 730 1660.
Sarlon Industries Pty Ltd, PO Box 130, Riverwood, NSW 2210. Enquiries: (02) 796 2222 (Sarlon sell weed mats.)
Seagrow Products, RSD Sunday Road, Callignee, Vic. 3844. Enquiries: Gerhard Grasser (proprietor, and very helpful), (051) 95 5396
Tom Garner (Ezi-Wool Mats), PO Box 50, Ringwood, Vic. 3134. Enquiries: (03) 876 2723
Universal Recycling (Garden Energiser), Berends Drive, Dandenong, Vic. 3175. Enquiries: (03) 706 4995.

CHIPPER/MULCHERS

These companies do not necessarily supply machines direct to customers, some sell their products through retail outlets such as hardware shops. However, they all supply information.
Allpower Industries Australia Ltd, PO Box 358, Box Hill, Vic. 3128. Enquiries: (03) 890 3344

Fauser Worms, Reply Paid 46, PO Box 400, Eastwood, SA 5063. Enquiries: (008) 88 2811

Motorcolt Australia Pty Ltd, Cnr Abbots Road and England Street, Dandenong, Vic. 3175. Enquiries: (03) 794 6222

Rover Mowers Ltd (Head office and factory) 155 Fison Avenue West, Eagle Farm, Brisbane, Qld 4007. Enquiries: (07) 868 0222

USEFUL PLANTS

Exotic Groves, PO Box 125, Innisfail, Qld 4860.

Honeysuckle Farm, Mountain Highway, Sassafras, Vic. 3787.

Meadow Herbs Cottage Nursery, PO Box 57, Mt Barker, SA 5251.

Paradise Distributors, 27a Bonney St, Nambour, Qld 4560. Enquiries: (074) 41 5921.

The Botanic Ark, Cnr Sutton and Copeland Rds, Warragul, Vic. 3820. Enquiries: (056) 23 5268.

The Fragrant Garden, Portsmouth Rd, Erina, NSW 2250.

USEFUL SEEDS

Digger's Seeds (including a special clover mix for mulching), 105 Latrobe Parade, Dromana, Vic. 3936.

Fruit Spirit Botanical Gardens, Dorroughby, NSW 2480.

Green Harvest, 52E Crystal Waters MS16, Maleny, Qld 4552.

New Gippsland Seed Farm, PO Box 1, Silvan, Vic. 3795.

Phoenix Seeds, PO Box 9, Stanley, Tas. 7331.

The Seed Savers Network, PO Box 105, Nimbin, NSW 2480.

Index

Another bestseller for the environmental gardener . . .

EARTHWORMS IN AUSTRALIA

A Blueprint for a Better Environment

David Murphy

'This can only be described as a great book, desperately needed . . . the knowledge and enthusiasm of the author comes through so strongly, it is difficult to put down' — Rosemary Davies & Peter Cundall, *The Weekly Times*

'The worm "bible" you've been waiting for' — Scott Kelleher, *Gardening Australia*

'One of the great titles of the year' — *The Sunday Mail*

'Everyone contemplating worm composting or worm farming should use this book as a reference' — Allen Gilbert, *Australian Horticulture*

'The first book I have ever read dealing solely with earthworms . . . fascinating' — *Ballarat Courier*

Using the very latest information from the world's leading agricultural research centres (including Australia's own CSIRO), *Earthworms in Australia* provides the first comprehensive and easily understood guide published for Australian gardeners, farmers, conservationists and aspiring worm farmers.

Fully illustrated, with forewords from four of this country's top agricultural scientists, this breakthrough book clearly shows us that the answers to our tough environmental questions have been under our feet all along!

ISBN 1 875657 09 6 Paperback $14.95

Also by Sandra Clayton . . .

THE REVERSE GARBAGE GARDEN

Written and illustrated by Sandra Clayton

'This is a timeless book for now and for future generations, the book we had to have to cope with the stress of modern day living, soil degradation, poverty and environmental pollution' — Allen Gilbert

'Essential reading for all those concerned with the environment . . .' — *Pools & Gardens*

'For the environmentally conscious gardener this book is a must' — *The West Australian*

'This book is so well written and illustrated . . . *The Reverse Garbage Garden* will influence people to change their lifestyles' — *Ballarat Courier*

'Recommended reading for all' — *Geelong Advertiser*

Based on years of experience and written in a friendly, anecdotal manner, *The Reverse Garbage Garden* is packed with illustrations, charts, diagrams, and colour photographs designed to make it easy to be green and save money in the process.

Sandra Clayton's supremely useful and universally celebrated bestseller can empower anyone, whether they have a window box or a farm, to safeguard the environment and harvest highly nutritious organic fruit and vegetables. Includes a planting guide for each of Australia's climatic regions.

ISBN 1 875657 12 6 Paperback $19.95

Also published by Hyland House . . .

GROWING INTO GARDENING

with your children

Melissa McGroarty & Noelle Weatherly

'There are few well written books to introduce children to the joys of gardening, but *Growing into Gardening* fills the role perfectly' — Graeme Purdy, *The Herald-Sun*

'An excellent book written in an accessible, straightforward manner . . . Children are taught how and why plants grow and how they can be environmentally aware in the garden' — Cheryl Maddocks, *The Australian*

Packed with cartoons, illustrations and plenty of friendly advice and information, *Growing into Gardening* features activities, games and fact sheets that encourage an understanding of nature's ways, and also make for a lot of fun!

Published in association with the Royal Horticultural Society of Victoria.
ISBN 1 875657 19 3 Paperback $19.95

PRUNING ORNAMENTAL SHRUBS & TREES

An ABC for Australian Gardeners

L. F. Eager

'A brilliant little book, listing 270 species of ornamental shrubs and trees most commonly grown in Australian gardens and detailing when and how to prune them' — *The Sun-Herald*

'Compulsory reading for all those who take a fiendish delight in administering a short back & sides to their plants each year' — *Geelong Advertiser*

'Exceptional value for new and experienced gardeners alike' — Rob Pelletier, *Nunawading Gazette*

ISBN 0 947062 74 2 Paperback $14.95

CHAMOMILE FARM

Lyle Williams

'Lyle Williams' delightful book tells the story of Chamomile Farm from its uncertain beginning as a paddock of turnip weed to its present position as a showplace of successful organic gardening' — *Grass Roots*

'It is a gardening book with a difference — a human interest story told with feeling of a man's family and the way they have built a successful enterprise' — *Weekly Times*

ISBN 0 947062 08 4 Paperback $19.95